Nanotechnology
Health and Environmental Risks

Second Edition

Nanotechnology
Health and Environmental Risks

Second Edition

Jo Anne Shatkin

CRC Press
Taylor & Francis Group
Boca Raton London New York

CRC Press is an imprint of the
Taylor & Francis Group, an **informa** business

CRC Press
Taylor & Francis Group
6000 Broken Sound Parkway NW, Suite 300
Boca Raton, FL 33487-2742

© 2013 by Taylor & Francis Group, LLC
CRC Press is an imprint of Taylor & Francis Group, an Informa business

No claim to original U.S. Government works

Printed in the United States of America on acid-free paper
Version Date: 2012928

International Standard Book Number: 978-1-4398-8175-0 (Paperback)

Library of Congress Cataloging-in-Publication Data

Nanotechnology : health and environmental risks / [edited by] Jo Anne Shatkin. -- Second edition.
    pages cm. -- (Perspectives in nanotechnology)
    Includes bibliographical references and index.
    ISBN 978-1-4398-8175-0
    1. Nanotechnology--Risk assessment. 2. Nanotechnology--Health aspects. 3. Nanotechnology--Environmental aspects. I. Shatkin, Jo Anne, editor of compilation.

    T174.7.S52 2013
    620'.5--dc23   2012028126

Visit the Taylor & Francis Web site at
http://www.taylorandfrancis.com

and the CRC Press Web site at
http://www.crcpress.com

Contents

List of Figures

List of Tables

Preface

Richard Feynman is often credited with launching the field of nanotechnology in 1959 with his speech, "There's Plenty of Room at the Bottom." Since that time, nanotechnology has become a reality and it has flourished in the last decade. Engineered nanoscale materials have entered the marketplace through industrial and consumer products that take advantage of novel physical and chemical properties that appear in the 1–100-nanometer range. At this scale, a material may, for example, become magnetic, conduct an electrical current, or change color. The exciting new properties that emerge at the nanoscale allow scientists to engineer amazing products: carbon stronger than steel and drug-delivery systems that cure cancer cell by cell. Today, nanoparticles are being used in almost every sector of our economy, creating solutions in medical treatment, rocket design, solar and wind energy, clean water, cell phones, cameras, and a multitude of other consumer products.

To continue to capture the promise of nanotechnology, critical questions about the broader impacts of these transformative products on society and the environment need to be considered. It is essential to determine that the benefit derived from using nanotechnology-enabled products outweighs the potential for harm. This risk/benefit relationship is evaluated through the science of risk assessment, a discipline that uses scientific methods to determine the likelihood that one will be exposed to a nanomaterial, the amount to which one might be exposed, as well as the potential for the nanomaterial to be hazardous to human health and the environment. Given that the global market for nanotechnology-enabled products is estimated to be US$2.4 trillion by 2014,[*] the importance of assessing risk across the product life cycle—from laboratory to manufacturing plant, from consumer use to disposal—has also increased.

Early scientific efforts to examine the potential hazards of nanomaterials focused on their interaction with biological and ecological systems to determine if they disrupt human health and environmental equilibrium. The need to translate these results into the broader context of product manufacture and consumer use is becoming more widely recognized and appreciated. Nanotechnology risk assessment scientists and the federal agencies participating in the National Nanotechnology Initiative (Nano.gov) recognize the importance of risk assessment at critical junctures along the product pipeline and have, over the last several years, applied the Life-Cycle Analysis paradigm to nanomaterials in order to

---

[*] Lux Research. 2009. *Nanomaterials state of the market Q1 2009.* New York: Lux Research.

evaluate environmental, health, and safety research data. Life-cycle analysis is an essential component of science-based policy and risk management decisions. Its implementation is challenging, as it necessitates the flow of information across the product pipeline and among stakeholders as diverse as scientists, engineers, manufacturers, investors, insurers, public health and environmental advocates, policy makers, and Congress.

The challenges of this effort are daunting, yet meeting them is essential for nanotechnology to realize its potential as a sustainable industrial and business enterprise that will drive innovation in America. This revision of the 2008 first edition of *Nanotechnology Health and Environmental Risks* contributes significantly to this discussion by informing stakeholders about cutting-edge approaches to risk assessment and life cycle analysis, establishing a common vocabulary for multistakeholder communication, and providing the information necessary for science-based risk management.

Sally Tinkle
Coordinator for NNI EHS Research
National Nanotechnology Coordination Office

Acknowledgments

Many people contributed to this revised edition of the book and I am grateful for their help and support. First, I wish to thank the talented and organized staff at CRC/Taylor & Francis Publishing, especially my publisher Nora Konopka, for envisioning this project and inviting my participation in it, as well as series editor Gabor L. Hornyak and my series coauthors.

I am grateful to my coauthors, each a dear colleague and friend: J. Michael Davis for his contribution to this volume and to the environmental community in developing Comprehensive Environmental Assessment, an area I have enjoyed collaborating with him on over these years; Rick Pleus, for shedding light on the state of toxicology science; Thomas Peters for sharing his cutting-edge work on occupational exposures to nanoparticles; and Ann Bostrom and Ragnar Löfstedt for expanding the scope of this book and sharing insights into risk perceptions and the mental models people use to evaluate technological risks, and for their expert contributions and friendship. Thanks especially to Sally Tinkle for her work to advance the concepts of life cycle risk assessment and for contributing the preface.

I wish to thank numerous colleagues in the fields of environmental science and policy, colloidal chemistry, nanoscience, and risk analysis, too many to name individually, who shared their perspectives and with whom I have had many insightful conversations about managing nanomaterials, nanotechnology, and other emerging environmental risks. Michael Ritchey of Boeing deserves recognition here because the course I developed at his request stimulated the development of much of the material in the book. I also acknowledge those contributors to the first edition of this book who made the revision so much easier: Paul Susman, Clayton Teague, Brenda Barry, Mike Davis, Tor Arnesen, and of course my dear sister, Susan Shatkin, who carefully edited and made that edition readable. I am also grateful to my dedicated high school science teachers, Mr. Vibert and Ms. Erlenmeyer, for introducing me to science and inspiring me to become a scientist (this would no doubt surprise them!). My perspectives on these issues have been heavily influenced by the thought leaders of the science technology and policy realm who developed new, interdisciplinary paradigms from their disciplines including: Chris Hohenemser, Jeanne and Roger Kasperson, and Ortwin Renn, as well as several mentors: Ron Levy, Halina Brown, Rob Goble, Susan Vick, George Hallberg, Katherine Hammond, John Kelly, and Bill Coleman.

I also wish to express my gratitude to my family for wholeheartedly supporting me in this endeavor. I thank Chris first and foremost, and

Josh and Tony, as well as Evelyn and Susan, for their love and encouragement and their enduring belief that I can accomplish whatever I choose to. And finally, I am grateful for my beloved friends and extended family for understanding when I couldn't show up.

Editor

Jo Anne Shatkin, PhD, leads CLF Ventures, a nonprofit affiliate of the Conservation Law Foundation, New England's most influential environmental advocacy organization. CLF Ventures develops and assists in the implementation of innovative market solutions to pressing environmental problems. Dr. Shatkin is a recognized expert in human health risk assessment, emerging contaminants policy and environmental aspects of nanotechnology. She provides leadership on proactive approaches to sustainable technology development, working to advance life cycle approaches to risk analysis, and incorporate life cycle thinking into product design and development. Using screening tools such as LC Scan, a life cycle scanning tool for product developers, and NANO LCRA, a life cycle risk assessment framework, Dr. Shatkin leads the CLF Ventures team in efforts to minimize impacts of products and technologies.

Since 2005, Dr. Shatkin has provided leadership on the responsible development of nanotechnology, and approaches for decision making under uncertainty. She teaches courses, and has published papers and book chapters on topics of environmental health and safety, and life cycle approaches to risk analysis for nanotechnology. Dr. Shatkin serves as councilor of the Society for Risk Analysis (SRA), is a founding board member of the Stewardship Action Council, and leads the SAC Collaborative Partnerships initiative. She serves on the board of the Center for Environmental Policy at American University, and was a Switzer Environmental Fellow. Dr. Shatkin received an individually designed PhD in environmental health science and policy in 1994 and her MA in risk management and technology assessment in 1990, both from Clark University, Worcester, Massachusetts, and possesses a Bachelor of Science degree from Worcester Polytechnic University in molecular biology and biotechnology.

Contributors

Ann Bostrom, PhD, is professor of public administration in the Daniel J. Evans School of Public Administration at the University of Washington, Seattle. Her research is devoted to investigating how people understand and make decisions about environmental, health, and technology-related risks, with specific focus on their mental models of hazardous processes. She holds a BA in creative writing from the University of Washington, an MBA from Western Washington University, and a PhD in public policy analysis, Carnegie Mellon University, completed Fulbright studies with a Lois Roth Endowment scholarship at the University of Stockholm and conducted postdoctoral studies in engineering and public policy at Carnegie Mellon University and in cognitive aspects of survey methodology at the Bureau of Labor Statistics. Dr. Bostrom is the current president and a fellow of the Society for Risk Analysis.

J. Michael Davis, PhD, was a senior science advisor in the National Center for Environmental Assessment until his retirement in January 2012 from the U.S. Environmental Protection Agency after more than 30 years of service. He earned his PhD from Duke University in 1973 and held postdoctoral fellowships at the University of Oxford, England, and at the University of North Carolina at Chapel Hill. While at the EPA he conducted risk assessments dealing with fuels and fuel additives including lead, manganese, and MTBE. His major area of focus for the past several years has been applying Comprehensive Environmental Assessment (http://pubs.acs.org/doi/abs/10.1021/es3023072) to the identification and prioritization of research directions for nanomaterials. He is currently an independent consultant.

Thomas M. Peters, PhD, is an associate professor and director of the Industrial Hygiene Program at The University of Iowa, Department of Occupational and Environmental Health. He develops novel sampling methods that he then applies to understand and control aerosols in the workplace and the environment. In recent projects, Dr. Peters has developed methods to assess airborne engineered nanomaterials apart from background aerosols through activity monitoring with direct-read instruments and computer-controlled single-particle electron microscopy of collected particles. In addition, Dr. Peters has developed passive sampling techniques to investigate the variability in composition of particles in the atmosphere. He has published numerous papers on the subject of assessing exposures to nanoparticles in workplaces and has recently developed a new chapter entitled "Engineered Nanomaterials" in Patty's Industrial Hygiene.

Ragnar E. Löfstedt, PhD, is professor of risk management and the director of King's Centre for Risk Management, King's College London, UK where he teaches and conducts research on risk communication and management. He is also an adjunct faculty at the Harvard Center for Risk Analysis, Harvard School of Public Health where he directs the Risk Communication Challenge Course for continuing education professionals; adjunct professor at the Department of Engineering and Public Policy, Carnegie Mellon University; and a visiting professor at the Centre for Public Sector Research, Gothenburg University, Sweden. He has conducted research in risk communication and management in such areas as renewable energy policy, food safety issues, pharmaceutical recalls, telecommunications, biosafety, and the siting of building of incinerators, nuclear waste installations, and railways. Dr. Lofstedt is the author/editor of ten books and more than ninety peer reviewed articles; editor-in-chief of *Journal of Risk Research*; editor of the Earthscan publications' Risk, Society and Policy book series, and serves on the editorial boards of *Journal of Health Communication, International Journal of Risk Assessment and Management,* and *Risk Management.*

Richard C. Pleus, PhD, is the founder and managing director of Intertox, Inc., an independent scientific consulting and research organization. Dr. Pleus is also the cofounder and chief scientist of Intertox Decisions Sciences, LLC, a risk management company offering software and database solutions for several industries, including nanotechnology. He has more than twenty-five years' experience as a toxicologist assessing the risk to humans exposed to chemical and biological agents via water, air, soil, therapeutic agents, and consumer products. He is a U.S. delegate on the International Organization for Standardization (ISO) Technical Committee (TC) 229, Nanotechnologies, and led the U.S. Technical Advisory Group (TAG) Working Group 3 to develop a comprehensive list of physical and chemical characterization parameters of engineered nano-objects for toxicologic assessment. He is also a U.S. delegate for the U.S.–Russia Bilateral Presidential Commission of Science and Technology, chosen for his expertise on nano-related ESH issues. Dr. Pleus is also the cofounder of the Nanotechnology Health and Safety Forum and is a scientific advisor to the NanoSafety Consortium for Carbon. Dr. Pleus's credentials include a BS in physiology, with honors, from Michigan State University; an MS in environmental Health and a PhD. in environmental toxicology from the University of Minnesota; and postdoctoral research in neuropharmacology at the University of Nebraska Medical Center.

1

Introduction: Assessing Nanotechnology Health and Environmental Risks

Jo Anne Shatkin

CONTENTS

Exposure to free engineered nanomaterials (as opposed to fine particles that are naturally occurring or that are the incidental by-products of human activities such as combustion or welding) is for the most part still low. So we are well positioned to assess possible risks before nanoparticles become widely used or make their way into the environment in large quantities.

E. Clayton Teague
Former Director, U.S. National Nanotechnology Coordination Office

You've probably picked up this book because you want to know, "Is nanotechnology good or bad? Should I adopt it? If I have already adopted it, what do I need to know about it? Does nanotechnology answer the pressing problems of our day?" Or maybe you are wondering what nanotechnology is and whether it is something you should worry about.

It is widely agreed that the opportunities for nanotechnology development are vast and represent enormous potential for technological innovation to devise smarter, more precise solutions to meet a breadth of human needs. The ability to engineer at the nanoscale means technology can be designed and developed to specifically address societal needs, and this specificity presents an opportunity for sustainable technology development, but only if the goals are established and widely agreed upon.

Throughout my career, I have been amazed at the tiny risks that people are really afraid of, and equally amazed at the large risks taken daily—sometimes by the same people—that are so well known and so dangerous. Risk analysis is a way to put different hazards in perspective, to ensure focus on the most significant concerns, which are not necessarily the ones people are most vocal about. According to some projections for nanotechnology, everyone may be affected one way or another as nanomaterials and nanotechnology increasingly enter more sectors of the economy. If the projections are correct, then it is important to gain perspective on the potential risks of nanotechnology to inform your decision making—whether you are a developer, user, investor, consumer, or regulator, or simply curious. This book explores the respective fields of risk analysis and nanotechnology, and proposes an adaptive framework for taking on the challenge of risk assessment of this rapidly developing field of technology.

Many of the current applications of nanotechnology are in consumer products. Maybe the touch screen on your smartphone or MP3 player uses nanotechnology. Your tablet may use nanotechnology for a stronger, scratch-resistant coating or a more energy-efficient processor or higher resolution display. You may be applying paint with nanomaterials in it. You could be wearing antimicrobial socks or static-free pants with nanotechnology, or your baby is using a nano-coated blanket. Your home may have a self-cleaning toilet (or perhaps you wish it did). You may be working with nanomaterials in a laboratory.

This book explores questions about the introduction and uses of nanotechnology for energy, industry, medicine, technology, and consumer applications. It also looks into the issue of how to determine whether there is risk—and ways to manage the risk—even if there is little in the way of reliable evidence. Why is there so much attention to the risks associated with nanotechnology? Why write a book—or read one—about tiny, tiny risks that may or may not be more than theoretical? Considered in the context of global environmental and public health concerns such as climate change, natural disasters, war, AIDS, malaria, antibiotic resistance, or nuclear threats, how significant are the potential risks from nanoscale materials and nanotechnology? What are the risks from nanotechnology, and why should we even consider them? The number one answer is that nanomaterials and nanotechnologies offer benefits beyond the ways we currently treat disease, produce energy, manufacture products, and attend to our daily wants and needs. But, as with any emerging technology, we must identify and address the potential, yet uncertain, risks.

The unique behavior of substances engineered to sizes 100 nanometers and smaller (a nanometer is one billionth of a meter) is what makes them attractive to nanoscientists and engineers for developing new materials and applications in almost every sector of the world's economy. The size raises questions, however, about whether their unique behavior also affects biological systems (people and the environment) differently than other materials.

Deoxyribonucleic acid (DNA), the basic building block of life, is made of molecules that are only a few nanometers in width. Because it occurs in nature, DNA may not fall into everyone's definition of nanoscale materials; however, a key point is that other nanoparticles and nanoscale materials are in the same size range as DNA, and perhaps these can react with one another in ways that larger particles cannot. Simply, it is important to address the questions of health and environmental risks now, so that nanomaterials and nanotechnologies can fulfill their promise to improve medical diagnoses and treatment, help address and reduce the impacts of our energy needs and mitigate global warming, help reduce existing pollution, and perhaps even provide tools to stem global pandemics. Answering the questions about health and environmental risks from nanotechnologies raises broader societal implications about managing any emerging technology, about how to address the societal implications of technological change, and about how to make decisions about them under uncertainty.

Evaluating the potential risks early in product development creates an opportunity to safely manage them. The main reason to address the health and environmental aspects of nanotechnology is uncertainty. Even while the behavior of nanomaterials in biological systems is poorly understood, there are well over 1,000 products containing nanomaterials on the market today and many more in the pipeline, including more sophisticated applications that include active nanostructures that change in response to an external stimulus. Some people are concerned that the uncertainties associated with the introduction of nanotechnology may create new and unmanageable hazards to health and to the environment.

There is no reason for technology to develop in an unsustainable manner. In the past, lack of foresight has yielded staggering costs in terms of lives and lost use of land—costs to corporations, governments, and individuals—that could have been avoided by proactive efforts to manage risks. The tools to develop safer technologies and less harmful products exist. There is also plenty of experience that demonstrates how *not* to proceed. The chemical industry has learned this, and now participates in voluntary and regulatory efforts to create safer products (e.g., HPV Challenge, Responsible Care). The electronics industry (IEEE and IEC) provides leadership and voluntary initiatives to develop nanotechnology safely by developing standards. Those developing nanoscale materials and using them in technologies have a responsibility to ensure that they do not harm people's health or the environment. Governments and many industries recognize this and have initiated efforts to do so, but the complexities raised by some of the unique properties of nanomaterials studied to date raise concerns about how to measure and assess the risks of nanomaterials in products, and there is still uncertainty regarding how to manage those risks.

The field of risk analysis has grown enormously in the past four decades and provides a systematic, coherent, and tested foundation for managing the uncertain health and environmental aspects of nanotechnology. Years of managing hazardous substances has led to development of sophisticated tools for evaluating the behavior of materials, and by identifying the needs early, industry and governments can conduct the necessary research and make better decisions about how to manage nanomaterials to ensure a safer path forward. Risk analysis offers the tools to identify and manage risks under uncertainty; however, the challenge is to adapt it to the rapid stream of developments in nanotechnology that may affect environmental and health hazards.

Nanotechnology is a major player in the technological future, and there is an exciting opportunity to design that future. Nanotechnology presents an opportunity to redesign and to engineer technologies to specification. This offers the chance to minimize the risks and maximize the benefits of technological innovation. So if it seems a material or technology may be more hazardous than we are willing to accept, this is an opportunity for innovation—to engineer out the hazard. New materials can be (and increasingly are) designed to be safer (so-called green manufacturing) and more environmentally friendly. Risk analysis is a tool to help achieve a sustainable future with nanotechnology.

Many people within and outside of government are working to ensure the safety of nanomaterials, to avoid the unintended effects that occurred from other substances: asbestos, polychlorinated biphenyls (PCBs), the insecticide DDT, and lead, for example. These substances were widely used because they offered solutions to many industrial and societal needs, but over decades of use, their impacts on health and the environment began to emerge. While the health and environmental effects of these substances are among the most studied, they are still poorly understood, suggesting that a different path is needed for identifying and managing substances and technologies going forward.

This book seeks to give the reader some tools and information to help improve understanding of the health and environmental dimensions of nanotechnology and make better decisions about using them. As with other books in the series, this volume is intended to be accessible to the nonspecialist and explores a breadth of technical and societal topics relevant to the discussion. This book cannot answer most questions about the health and environmental risks of nanotechnology; the answers simply are not known today. But it does provide perspectives about what types of risks could exist, and what can be done to address them.

1.1 What Is Nanotechnology?

Nanotechnology is a *scale* of technology, not a type, and it has applications in every economic sector: medicine, energy, industrial applications, materials science, engineering, electronics, communications, packaging, cosmetics, additives, coatings, food science, water purification, and agriculture. Nanotechnology is also an enabling technology, that is, the unique properties at the nanoscale can be harnessed for virtually any technological application. Hence, we ought to consider them "nanotechnologies" rather than nanotechnology. Ever Google "nano"?

In 2007 there were 61 million hits, in 2012 there are 320 million—many of which have nothing to do with nanotechnology, but one finds "nano" everywhere—some of the discussion is hype about the promise or the perils (for example, an advertisement for a "nano" car, the Tata), and it is critically important to consider the source of information. But there are also numerous applications in development that suggest a brave new technological world—self-assembling materials that act on demand—not the world of *Prey*, a science fiction novel in which intelligent nanoscale robots take over (Creighton 2002), but of smart technology that combines information technology with nanotechnology in novel applications. An example might be a targeted pesticide, released only when a specific pest appears, triggered by detecting a key protein, rather than routine spraying of crops.

Nanomaterials have been in commerce for decades, including carbon nanotubes, first patented in the 1990s, but the pace seems to be accelerating now. One database of consumer products containing nanomaterials or using nanotechnology lists over 1,300 entries of products on the market today (WWCS 2012), and there are numerous industrial applications that use nanotechnology or incorporate nanomaterials into products. Funding for research and development of nanotechnology is also rapidly accelerating. In the United States in 2007, there were over 12,000 patents containing the term *nano*. In 2012, the term *nano* is used in over 32,000 patents. The number of nanotechnology patents worldwide doubled between 1998 and 2003 (Hullman 2006) and has nearly tripled in the last five years. Most major universities now offer programs in nanoscience, or at least have researchers working in it. In 2003, over 1,000 university and research institutions and nearly 1,200 private companies were actively working in nanotechnology. There are over 2,500 organizations registered in the European database *Nanoforum* (www.nanoforum.org), up from 1,500 organizations in 2005. Over 1,000 of these are commercial enterprises in Europe alone! There has been significant growth in Asia, Russia, and South America as well as in the developed economies of North America, Japan, Europe, and Australia. Mikhail Roco, who envisioned and worked extensively to create the U.S. National Nanotechnology Initiative, sees vast and far-reaching implications. "Besides products, tools and healthcare, nanotechnology also implies learning, imagination, infrastructure, inventions, public acceptance, culture, anticipatory laws, and architecture of other factors" (Roco 2004). In reflecting on ten years of investment by the U.S. government, Roco and colleagues highlight the record size of the semiconductor industry ($300 billion in 2010) attributing 60% of that to nano-enabled products (Roco, Mirkin, and Hersam 2011).

As you can begin to appreciate, nanotechnology is a vast, complex field that is rapidly developing in all directions. My aim is to make this discussion accessible and understandable to an audience that includes those unfamiliar with nanotechnology or risk analysis. I especially want to avoid a lot of dense scientific terminology. However, it is necessary to offer some definitions to clarify the discussion. This is particularly important because, with discussions about nanotechnology occurring in so many fields, an interdisciplinary conversation has begun. As scientists explore behavior at the nanoscale, phenomena never seen before may complicate learning about the behavior of nanotechnology. That is to say, the tools and instruments that allow observation at the nanoscale will lead to discoveries unrelated to nanotechnology *per se*, i.e., that simply relate to the ability to see things not previously observable. As it turns out, how nanotechnology is defined affects how the risks from nanotechnology are addressed.

1.1.1 What, Then, Is Nanotechnology?

Nano is a prefix used in the metric scale to represent one billionth. A nanometer (or nm) is one billionth of a meter (≈39 in.). Nano comes from the Greek word for *dwarf*, so in combination with technology, it becomes dwarf technology. Technology applies science and materials for human uses, and nanotechnology applies science and materials at the nanoscale. People refer to nanotechnology as "tiny tech" or "nanotech": It represents the scaling down of technology to a new scale, generally agreed to be in the range of 1 to 100 nm.

A nanometer is so small, it is hard to conceptualize. The head of a pin is 1 million nanometers wide. A strand of hair is 100,000 nm wide; the size of most bacteria is roughly 1,000 to 5,000 nm; viruses are about 100 to 500 nm in width. DNA, the genetic foundation for life, is about 1 to 2 nm wide. Nanotechnology is on the scale of our DNA. However, not everything at the nanoscale is nanotechnology. Natural and human-generated nanoparticles occur in the environment. Food particles, like protein and sugars, are nanoscale. Air pollution, for example, includes nanoscale particles, but these are not technology. They are not manufactured or specifically designed to use nanoscale properties. Fires and volcanic eruptions release nanoparticles to the atmosphere, but these occur naturally, not the result of human activities, and are nanoscale particles, not nano*technology*. Smoke and water vapor (steam) also include nanoparticles.

According to the U.S. National Nanotechnology Initiative (NNI), "Nanotechnology is the understanding and control of matter at dimensions of between approximately one to one hundred nanometers where unique phenomena enable novel applications" (NNI 2012). An alternative definition from the American Society for Testing and Materials (ASTM) International is: "Nanotechnology is a term referring to a wide range of technologies that measure, manipulate and incorporate materials and/or features with at least one dimension between approximately 1 and 100 nanometers (nm)" (ASTM 2007). Such applications exploit the properties distinct from bulk/macroscopic systems of nanoscale components. Unfortunately, there is no consensus on how to define nanotechnology or nanomaterials, and dozens of different definitions of nanotechnology have been offered. Fortunately, most share two basic tenets: size and unique properties.

One of the main reasons for the explosion of interest in nanotechnology is the unique properties and behavior of matter at the nanoscale. When particles are synthesized at the nanoscale, their properties change. For one thing, *nanoparticles have much more surface area compared to their weight than larger particles.* This single property means that much less material can be used for an application, allowing us to save natural resources, energy, and money, provided that it does not cost more to produce. Using less material in products can offer both economic and environmental benefits.

The greater surface area relative to mass also means that nanoscale materials are more reactive than larger particles, so less goes farther. In addition, because of the size and/or structure, some basic chemical and physical properties change. Titanium dioxide, a white pigment widely used in coating materials and consumer products such as paint and toothpaste, becomes transparent when manufactured at the nanoscale. Nanoscale titanium dioxide is used, among other purposes, to make sunscreen and other ultraviolet ray–resistant coatings transparent. Gold changes color at the nanoscale—depending on the size of the particles, it can be orange or red—so it is being used in sensing technology, the color being an indicator of a particular reaction. It is not shiny or conductive, as macro-gold is. At the nanoscale, it behaves as a semiconductor.

Imagine a surface so smooth that water cannot stick to it, and instead rolls off. Imagine a sensor so sensitive that it can detect a single molecule of a contaminant in your drinking water. A few other current and proposed applications that use nanotechnology include:

Self-cleaning and air-purifying surface coatings: Such coatings incorporate nanoscale titanium dioxide into surface coatings for toilets, window glass, and building exterior and interior walls. Surfaces are smooth and, when exposed to light, cause chemical reactions, killing bacteria and sweeping away dirt, thereby creating a more sanitary environment, safer windshields, and walls that become graffiti-proof. This saves water and cleaning chemicals, and surfaces remain intact longer.

Reducing pollution: Applications range from removing nitrogen oxide from air pollution on the walls of buildings (by nanoparticles reacting with the air pollutants, breaking them down) to breaking down chlorinated solvents in groundwater.

Self-healing coatings: Such coatings are applied to the surfaces of automobiles and other equipment that can repair themselves after being scratched.

Stronger, more flexible sporting equipment: Such equipment includes golf clubs, bicycle frames, tennis balls, and baseball bats made from lighter composite materials, thereby improving performance.

Static and wrinkle-free fabrics: Fabric fibers are very small and thus do not wrinkle, preventing charge buildup.

Antibacterial applications: Such applications include bandages and wound covers, kitchen equipment, socks, underwear, camping gear, doorknobs, bus seats, waiting room furniture, and medical devices that kill bacteria on contact.

Water filtration: This application includes devices that regenerate and continually remove bacteria and chemicals from drinking water.

These products are on the market today. In the gee-whiz category, the following applications are in various stages of research and development, and some may soon become commercial products:

Smart food packaging: This could lead to a food container that tells you if the food inside has gone bad. (Some have suggested packages that release substances to treat and purify contaminated foods.)

Electrically conducting thin films: An electronic surface could be made so thin that it becomes more like fabric, but can be used for electronic displays of video, cell phones, and computer screens. Imagine your clothing being a video display, your mobile phone wrapped around your wrist, your MP3 built into your jacket, or your tablet folding up in your pocket.

Solar paint: Solar cells could be so small they can be incorporated into paints and applied directly onto buildings.

Doctor on a chip: A medical diagnostic test could find a problem, diagnose it, and automatically treat it, for example, a cancer cell detector that selectively kills cancer cells.

Fuel cells: Fuel cells could be powered by viruses.

Light emitting diodes: A breadth of materials could improve resolution and lower energy usage.

Lighter, stronger composite materials: These could be used for automotive and aerospace applications, increasing safety and fuel efficiency. (Can you say "paper airplanes"?)

Smart dust: Tiny sensors, smaller than dust particles, could be used to monitor the environment, ensuring the air is pure, conducting surveillance, measuring pollen, checking for chemical weapons, air pollution, or wind speeds in real time.

Looking ahead, as with the advent of automation in the twentieth century, many of the proposed applications of nanotechnology could even further streamline the way people live. Some features might alleviate the need for manual surface cleaning. Nanoscale additives are used in some coatings that are so smooth that nothing can stick to it, for example: windshield treatments and aerospace coatings that become "self-cleaning" surfaces that could limit the need to dust, mop, sweep, wash windows, power-spray, clean toilets, and wash dishes. These examples focus on consumer applications, but there are many industrial benefits, including better energy transmission, improved water treatment technology, nonstick and antibacterial medical and food preparation surfaces, and many others.

1.1.2 Defining Nanotechnology and Nanomaterials

Typically, definitions of nanotechnology or engineered nanomaterials include two criteria: size (e.g., 1 to 100 nanometers [nm]) and something along the lines of "unique properties." It's worth noting that there is no universal definition for nanoscale materials or for nanotechnology; this is critical, because it determines what falls into the bucket, particularly in regulatory settings. As can be seen in Table 1.1, there are dozens of "official" definitions, and there is little agreement on how to define nanoscale materials within or across governmental and non-governmental organizations. That is, virtually no one uses the same definition. The International Standards Organization (ISO) defines nanotechnology as "understanding and control of matter and processes at the nanoscale, typically, but not exclusively, below 100 nanometres in one or more dimensions where the onset of size-dependent phenomena usually enables novel applications, where one nanometre is one thousand millionth of a metre" (ISO 2010). The U.S. Office of Pesticide Programs (OPP) defines a *nanoscale material* differently from most definitions by avoiding the terms *manufactured* or *engineered*. OPP defines nanoscale materials based on size and process ("intentionally produced") rather than size and unique properties (EPA 2011). A recent review of the data led researchers to suggest that engineered nanoscale materials should be defined by size-dependent properties, not simply size (Auffan et al. 2009). OPP's definition is much broader than others, since it includes all nanoscale materials that are intentionally produced. This emphasis away from properties, and on size alone, casts a broader net, excluding only nanoscale particles that are produced incidentally, during manufacturing. A sampling of definitions of nanomaterials is given in Table 1.2.

1.2 The Roots of Nanotechnology and the Next Industrial Revolution

A brief look back at the historical evolution of nanotechnology gives a deeper understanding of its only currently imagined future potential. Richard Feynman, the legendary Caltech physicist, gave a seminal talk at the American Physical Society in 1959 entitled, "Plenty of Room at the Bottom." Feynman described how technology can miniaturize and continue to miniaturize. He conceptualized shrinking technology from its current scale to a scale one-sixteenth its size, repeatedly, until achieving technology at the scale of matter. At that point, he theorized about the ability to manipulate matter at the atomic scale (Feynman 1959).

TABLE 1.1

Official Definitions of Nanomaterials and Nanotechnologies

Organization	Definition
Australia Commonwealth Scientific and Industrial Research Organisation	"Nanotechnologies are emerging fields that use extremely small particles, called nanomaterials, to produce new processes, devices and products for a variety of areas, including energy, health and water."
Australian National Industrial Chemicals Notification and Assessment Scheme	"Nanotechnology is engineering at the atomic or molecular level. It is a group of enabling technologies that involves the manipulation of matter at the nanoscale (generally accepted as 100 nanometres or less, where a nanometer is one-billionth of a metre) to create new materials, structures and devices."
Australian Office of Nanotechnology—National Nanotechnology Strategy	"Nanotechnology is the collective term for a range of technologies, techniques and processes that involve the manipulation of matter at the nanoscale—the size range from approximately 1 nanometre (nm = one millionth of a millimetre) to 100 nm."
Australian Pesticides and Veterinary Medicines Authority	"Nanotechnology is a process of deliberately engineering materials at the atomic or molecular level to create nanomaterials."
Australian Therapeutic Goods Authority	"The term nanotechnology is used to describe a wide range of methods involved in the production and engineering of structures and systems by controlling size and shape at the nanometre scale."
Food Standards Australia New Zealand	"While there is no internationally agreed definition for 'nanotechnology,' the term is usually applied to the process of controlling the size and shape of materials at the atomic and molecular scale. Generally, the term applies to deliberately engineered matter less than 100 nanometres (nm) in size in one dimension."
Safework Australia	"The precision-engineering of materials at the scale of 10^{-9} metres at which point unique properties occur."
ASTM	"A term referring to a wide range of technologies that measure, manipulate, or incorporate materials and/or features with at least one dimension between approximately 1 and 100 nanometers (nm). Such applications exploit the properties, distinct from bulk/macroscopic systems, of nanoscale components."
British Standards Institute	"Nanotechnology is an emergent area that is developing quickly and is the branch of science and engineering that studies and exploits the unique behaviour of materials at a scale of 1–100 nanometres."

(continued)

TABLE 1.1 (CONTINUED)

Official Definitions of Nanomaterials and Nanotechnologies

Organization	Definition
Health Canada	"Health Canada considers any manufactured product, material, substance, ingredient, device, system or structure to be nanomaterial if: a. It is at or within the nanoscale in at least one spatial dimension, or; b. It is smaller or larger than the nanoscale in all spatial dimensions and exhibits one or more nanoscale phenomena. For the purposes of this definition: The term 'nanoscale' means 1 to 100 nanometres, inclusive; The term 'nanoscale phenomena' means properties of the product, material, substance, ingredient, device, system or structure which are attributable to its size and distinguishable from the chemical or physical properties of individual atoms, individual molecules and bulk material; and, the term 'manufactured' includes engineering processes and control of matter and processes at the nanoscale."
ISO TC 229	"Understanding and control of matter and processes at the nanoscale, typically, but not exclusively, below 100 nanometres in one or more dimensions where the onset of size-dependent phenomena usually enables novel applications, where one nanometre is one thousand millionth of a metre."
OECD WPNM	"Nanotechnology is the set of technologies that enables the manipulation, study or exploitation of very small (typically less than 100 nanometres) structures and systems."
European Commission	"Nanotechnology is the study of phenomena and fine-tuning of materials at atomic, molecular and macromolecular scales, where properties differ significantly from those at a larger scale."
E.U. Nanoforum (2006)	"The manipulation or self-assembly of individual atoms, molecules, or molecular clusters into structures to create materials and devices with new or vastly different properties."
International Council on Nanotechnology	"Applications developed using materials that have at least one critical dimension on the nanometer length scale."
International Risk Governance Council	"Nanotechnology refers to the development and application of materials, devices and systems with fundamentally new properties and functions that derive from their small size structure (in the range of about 1 to 100 nanometers) and from the recent ability to work with and manipulate materials at this scale."

(continued)

TABLE 1.1 (CONTINUED)

Official Definitions of Nanomaterials and Nanotechnologies

Organization	Definition
BASF	"Nanotechnology is used to describe materials, structures and technologies involving the creation or presence of at least one spatial dimension smaller than a hundred nanometers (nm)."
ETC Group	"Nanotechnology [is] the manipulation of matter at the scale of atoms and molecules (a nanometer [nm] is one-billionth of a meter)."
NanobioRAIS	"Nanotechnology is really a combination of distinct technologies and sciences."
Food and Drink Federation	"Nanotechnology is an enabling technology with potential applications across a wide range of industrial and consumer products. It involves the manufacturing and utilising of the special properties of material at nanoscale (the nanoscale is regarded to range from about 1 nanometre [nm] to 100 nanometres)."
U.K. Soils Association	"Nanotechnology involves the manipulation of materials and the creation of structures and systems at the scale of atoms and molecules. This can be either through simple physical processes or by specific engineering. Nanoparticles are commonly defined as measuring less than 100 nm—one hundred millionths of a millimetre."
European Food Safety Authority (EFSA)	"Nanotechnology is a field of applied sciences and technologies involving the control of matter on the atomic and molecular scale, normally below 100 nanometers."
U.K. Royal Society	"Nanotechnologies are the design, characterisation, production and application of structures, devices and systems by controlling shape and size at nanometre scale."
U.K. Institute of Nanotechnology	"Science and technology where dimensions and tolerances in the range of 0.1 nanometer (nm) to 100 nm play a critical role."
U.S. National Nanotechnology Initiative	"The understanding and control of matter at dimensions of roughly 1 to 100 nanometers, where unique phenomena enable novel applications."
U.S. Environmental Protection Agency	"Research and technology development at the atomic, molecular, or macromolecular levels using a length scale of approximately one to one hundred nanometers in any dimension; the creation and use of structures, devices and systems that have novel properties and functions because of their small size; and the ability to control or manipulate matter on an atomic scale."

(continued)

TABLE 1.1 (CONTINUED)

Official Definitions of Nanomaterials and Nanotechnologies

Organization	Definition
U.S. Food and Drug Administration	Not defined

EFSA 2009. Text describes equivalence to engineered nanoparticle and to manufactured nanomaterials.

Eric Drexler (1986), an early pioneer in molecular manufacturing, began his famous book, *Engines of Creation,* with

> Coal and diamonds, sand and computer chips, cancer and healthy tissue: throughout history, variations in the arrangement of atoms have distinguished the cheap from the cherished, the diseased from the healthy. Arranged one way, atoms make up soil, air, and water; arranged another, they make up ripe strawberries. Arranged one way, they make up homes and fresh air; arranged another, they make up ash and smoke. Our ability to arrange atoms lies at the foundation of technology.

Of course, in 1959, at the time of Feynman's lecture, materials at the nanoscale were already in use in the chemical industry, among others, but it was not called *nanotechnology.* These applications were not manipulating matter or engineering to specification at the nanoscale—and that is the difference between chemistry and nanotechnology. The ability to design and manipulate matter at the nanoscale is what is unique about nanotechnology. Drexler (1986) explains: "Just as ordinary tools can build ordinary machines from parts, so molecular tools will bond molecules together to make tiny gears, motors, levers, and casings, and assemble them to make complex machines."

This is Drexler's vision for the future of nanotechnology. The manipulation of matter at the nanoscale and manufacturing on a molecule-by-molecule basis is called *molecular manufacturing.* At this point in time, it has not yet occurred. Many argue it is impossible, but the possibility of molecular manufacturing holds enormous potential benefits: safer products, the manufacture of air- and water-purification technology, and other environmental advantages such as making new materials and products without waste—no need for big stacks to treat, capture, or release air pollutants; no need for storm-water permits for releases to our water systems; and no need for trash and hazardous waste to go to our landfills or incinerators. And there is no need to waste the inputs to industrial processes. Drexler and many others who envision this future have paved the way for much creative work to come.

TABLE 1.2

Definitions of Nanomaterials

Organization	Term	Definition
U.S. FDA–CDER–Office of Pharmaceutical Science	Nanomaterial/ nanoscale material	"Any materials with at least one dimension smaller than 1,000 nm."
U.S. EPA Office of Pesticide Programs	Nanoscale materials	"An ingredient that contains particles that have been intentionally produced to have at least one dimension that measures between approximately 1 and 100 nanometers."
Friends of the Earth	Nanomaterials, n	"All particles up to 300 nm in size must be considered to be 'nanomaterials' for the purposes of health and environment assessment, given the early evidence that they pose similar health risks as particles less than 100 nm in size which have to date been defined as 'nano.'"
European Parliament	Nanomaterial	"An insoluble or bioresistant and intentionally manufactured material with one or more external dimensions, or an internal structure, on the scale from 1 to 100 nm."
International Standards Organization (ISO)	Nanoparticle, n	"Nano-object with all three external dimensions at the nanoscale that is the size range from approximately 1 nm to 100 nm."
ASTM	Nanoparticle, n	"In nanotechnology, a sub-classification of ultrafine particle with lengths in two or three dimensions greater than 0.001 micrometer (1 nanometer) and smaller than about 0.1 micrometer (100 nanometers) and which may or may not exhibit a size-related intensive property."
U.K. Soils Association	Manufactured nanoparticles	"Include: engineered nanoparticles that are intentionally produced to have a specific novel property,... and other manufactured nanoparticles that are produced incidentally by industrial processes, particularly modern high-energy processes such as those using high pressure (for example, some types of homogenisation)."

(continued)

TABLE 1.2 (CONTINUED)

Definitions of Nanomaterials

Organization	Term	Definition
Project on Emerging Technologies	Engineered nanomaterials	"A material that exhibits novel properties and behaviors (that cannot be predicted based on size alone) as a result of being manipulated at the nanoscale."
European Food Safety Authority	Engineered nanomaterials	"An engineered nanomaterial is any material that is deliberately created such that it is composed of discrete functional and structural parts, either internally or at the surface, many of which will have one or more dimensions of the order of 100 nm or less."
WHO–FAO Joint Panel on Nanotechnology in Food	Engineered nanomaterials	"Any material that is intentionally produced in the nanoscale to have specific properties or a specific composition."
EU Scientific Committee on Emerging and Newly Identified Health Risks (SCENIHR)	Engineered nanomaterials	"Any material that is deliberately created such that it is composed of discrete functional parts, either internally or at the surface, many of which will have one or more dimensions of the order of 100 nm or less."

This is the promise of nanotechnology—to transform the way energy is generated; how diseases are identified and treated; how food is grown; and the way fabrics, building materials, and consumer goods are manufactured—a promise based on current research as well as on applications that already exist in our economy, suggesting dramatic changes ahead. Many call nanotechnology "disruptive"—it will disrupt our current way of operating—radically changing our infrastructure, commerce, industry, trade, education, and manufacturing. It seems that nanotechnology represents the beginning of the next industrial revolution. Whether this is hype or an unrealized opportunity cannot be judged today. Yet the potential benefits of nanomaterials and nanotechnologies are transformative, and their impacts must be addressed in order to achieve their benefits. Other books in this series consider the societal dimensions of this disruption; the discussion here considers how nanotechnology may impact health and the environment.

1.3 Nanomaterials: The Current State of Nanotechnology Application

Drexler's vision of the future of nanotechnology through molecular manufacturing, though as yet largely unrealized, has been borne out thus far in the development of many new materials—"nanomaterials"—a prime example of which are those made from carbon. Carbon, the basic building block of life, can be formed into novel structures at the nanoscale, including nanotubes, nanowires, nanohorns, and the buckyball. Richard E. Smalley, with Robert Curl Jr. and Sir Harold Kroto, won the Nobel Prize in Chemistry in 1996 for their discovery of the C_{60} molecule they called buckminsterfullerene, after Buckminster Fuller, who developed the geodesic dome.

Buckminsterfullerenes are also called buckyballs or fullerenes, and most commonly contain 60 atoms of carbon in a spherical, soccer ball formation, with a diameter of about 1 to 2 nm (Figure 1.1 shows a C_{60} fullerene). Each of the intersections is a carbon atom; five or six atoms connect to form a ring; and the rings are interconnected, fused together in a cyclic molecule. Fullerenes can have more than 60 atoms of carbon (C_{60}); C_{70} is also commonly fabricated, and smaller molecules are possible, as are larger ones. There is a growing field of fullerene chemistry.

Fullerenes have been reported to behave as antioxidants, scavenging radical oxygen molecules. Hydroxyl radicals have been associated with aging and stress, and antioxidants are hot market items for skin creams and nutraceuticals. C_{60} may also disrupt cell membranes and reduce cell viability, and it is being investigated as an antibacterial additive for disinfectants. Other potential uses for fullerenes include industrial catalysts, drug-delivery systems, lubricants, coatings, electro-optical devices,

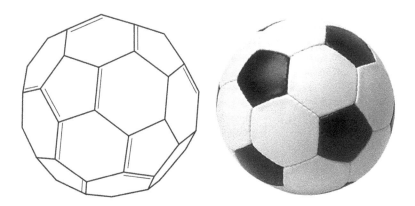

FIGURE 1.1
C_{60} fullerene.

and medical applications (e.g., antibiotics and targeted cancer therapy). However, the antioxidant properties of fullerenes could lead to their inclusion in other consumer products. C_{60} may also be used in lithium ion batteries, for fuel batteries, ultra-conducting material, highly functional paints, and industrial grinding materials.

In 2004, the first year of commercial-scale C_{60} manufacturing, production was estimated at 1,500 metric tons, but as uses develop, production costs are likely to decrease, and production volumes could increase. Near-future uses of fullerenes may be relevant to water, waste, air, energy, transportation, and pesticide applications. Its catalytic and antimicrobial properties may warrant use of C_{60} for water treatment and disinfection (Boyd et al. 2005) or as additions to products. Fullerenes could also have environmental applications in sensor technology, being able to measure other substances accurately and at low levels. C_{60} has also been measured in ambient air pollution from combustion sources such as vehicle exhaust (Utsunomiya et al. 2002), so even before it could be manufactured, C_{60} was part of our environment. However, the ability to manufacture C_{60} allows for new applications.

Carbon structures can become quite complex. One of the more common structures is the carbon nanotube. The formation of carbon nanotubes is like creating a roll of chicken wire. Laid flat, sheets of graphite (used in pencil lead) resemble chicken wire made of carbon. Rolled up into a tube, these become carbon nanotubes (see Figure 1.2). Nanotubes can be of variable length and width, and can be single-walled or multiwalled (multiple concentric tubes encasing one another). Each variation in structure brings new properties that at the time of this writing do not even have a standard nomenclature. That is, there is no universal classification for describing the length, width, ends, and characteristic properties of carbon nanomaterials and nanostructures. Any substitution of these molecules, or embedding a different molecule in them, also changes their structure and function. How many potential combinations are there? It has been estimated there are upwards of 50,000 combinations of carbon nanotubes (Colvin 2006).

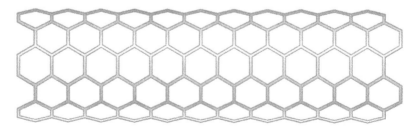

FIGURE 1.2
Carbon nanotubes.

What's so great about carbon nanotubes? They possess a breadth of electrical, optical, thermal, and physical properties that are being investigated for numerous applications. Some nanotubes conduct electricity better than copper, and depending on their charge properties, they can form an electrostatic coating for easier painting of surfaces. Nanotubes are stronger, lighter, and more flexible than steel, and they conduct electricity at low temperatures. Many current applications include adding nanotubes to composite structures to make them both stronger and more flexible. Examples include structural applications in aviation and automotive applications as well as sporting gear such as baseball bats, bicycle frames, golf clubs, hockey sticks, and tennis rackets. Carbon nanotubes also have applications in computing and electronics, for flexible displays, touch screens, circuits, memory chips, semiconductors, and conducting films with potential fuel cell and solar technology applications. The insulating properties of carbon nanotubes are being tested for flame-retardant applications. Some carbon nanotubes possess catalytic properties, that is, they are reactive with certain other substances. They can be used to decontaminate air and water, and they show potential for use in desalinization (removing salt from sea water to make it drinkable). In other words, nanotubes possess unique properties with advantages in most industrial sectors, and they could widely enter the economy in numerous applications.

Nanotubes and other nanostructures do not need to be made from carbon. Titanium, silica, and copper have all been used to make nanotubes. Many materials take on new properties at the nanoscale, and new structural configurations are likely. These materials, while novel by today's standards and currently in development in laboratories for numerous applications, are in actuality simple materials compared to the full-scale molecular manufacturing envisioned by Drexler (1986), who wrote that "our spacecraft are still crude, our computers are still stupid, and the molecules in our tissues still slide into disorder, first destroying health, then life itself." Drexler envisions biologically based technology, suggesting a level of complexity for the next generation of nanotechnology that few can currently envision. Nanotubes represent the future—engineered structures with specific properties used in a variety of applications.

1.4 Nanotechnology Risks Now

The commercialization of nanotechnology is literally under the microscope. Some nongovernmental organizations that focus on environmental health and consumer issues currently are calling for a moratorium on all products containing nanotechnology until their safety and risks are known (e.g., ETC Group, CTFA), while others are calling for transparency. The regulatory environment is dynamic, as agencies set policies, guidance, and new rules to address the new nano-enabled products entering commerce. Many of the international organizations discussed in Chapter 10 are working toward regulatory and voluntary standards for environmental health and safety of nanotechnology.

We are now at a point in history where consumers expect transparency, and safety, as well as sustainability of the products they use. Demands to understand and address risks in real time, that is, during development, add a difficult dimension to nanotechnology development. As will be discussed, the understanding of material behavior at the nanoscale is in its very early stages, and it is premature to make long-lasting decisions about applications of nanotechnologies without this understanding. However, now is the time to begin the analysis—while the actual risks from nanomaterials are small because they are produced in low levels and very few people are exposed in very small amounts—to guide decision making for when they are in widespread use. Addressing potential exposures now is the best way to mitigate the long-term risks of nanomaterials and nanotechnology, which are currently poorly understood.

1.5 Environmental Aspects of Nanotechnology

Many applications of nanotechnology benefit the environment, for example, treating drinking water, eliminating toxic chemicals, increasing water and energy efficiency, and harnessing cleaner energy technologies. How can the applications of nanoscience affect the environment? It is not clear today what the potential impacts are from nanoscale materials in the air, water, and soil. For example, it is not known to what extent nanomaterials might enter the food supply and become part of the human diet, or whether and how they can affect forests, coral reefs, or air quality.

Will there be a nano-environmental legacy? Are nanomaterials already entering the environment in ways that will allow them to persist and enter or upset the food chain? Will nanomaterials follow the path of other legacy pollutants, such as lead? How will this be determined if data are not being collected? One could argue that the amounts will be small, and in the near future, it is true that there are few applications of nanotechnology likely to allow free nanoparticles to enter the environment in significant amounts. However, as more and more applications adopt nanotechnology, the production, uses, and releases of nanoparticles will dramatically increase.

By way of example, in a hospital environment, it is very important to keep surfaces sanitary and free from contamination, and many cleaning and disinfection chemicals are used for cleaning equipment or washing floors and surfaces to help prevent the spread of germs. Using a product containing a nanomaterial as a disinfectant might mean it would be sprayed, wiped, poured into buckets and on floors, and washed down drains. An obvious question arises: Where could the nanomaterial end up? Any time chemicals are washed away with water or flushed down the drain, they are released into the environment. From drain pipes, these materials enter the groundwater and eventually can move to the nearest rivers and streams. Of course, this may affect drinking water sources and oceans.

Triclosan, commonly found in antimicrobial soaps and cleaning products, is among many consumer-use chemicals found in rivers and drinking water sources. Some populations of bacteria routinely exposed to substances designed to eradicate them (e.g., pesticides and medical antibiotics) are now found in the environment and have become resistant to antibiotics used in agriculture and to treat human diseases. Antimicrobial resistance is a big problem because bacteria are no longer susceptible to the treatments developed to kill them, and outbreaks can occur that cannot be managed. Currently, a number of hospitals are battling antibiotic-resistant *staphylococcal* infections (methicillin-resistant *Staphylococcus aureus* [MRSA]) in patients. More questions arise: If a nanomaterial is used in an antimicrobial treatment, can it cause antimicrobial resistance in the environment? What other unintended effects could a substance that is released in water cause?

One of many pathways that nanomaterials can enter the environment is through the drain pipes from a hospital that lead to a treatment plant where the water is treated and then released to the environment. (Figure 1.3 shows other pathways for nanomaterials to move in the environment.) What types of effects can occur from these environmental exposures? Nanomaterials could contaminate the water that is home to many plants and animals. Fish might absorb them from the water, or they could be taken up by bacteria and transformed to something else that is more toxic, more mobile, or more persistent. Nanomaterials could enter the food web. It is not always easy to predict what will happen when introducing materials into an ecosystem. Since all organisms require water to survive, this discussion focuses on the aquatic ecosystem, but there are many other environments to consider, including forests, deserts, mountains, tundra, savannahs, and broader marine systems, not to mention the ecosystem of the built environment of buildings and cities. By way of beginning the discussion, let us consider an example from the past that has structured much of the current framework of risk assessment and its application to nanotechnology.

FIGURE 1.3
Pathways for nanomaterials to move in the environment.

1.6 DDT, Learning from the Past

More than fifty years ago, a scientist named Rachel Carson motivated much of society's current environmental management by writing a book about chemicals in the environment and how the pesticide dichloro-diphenyl-trichloroethane (DDT) was harming birds and other wildlife. In *Silent Spring*, Dr. Carson explained how DDT and other chemicals were entering the environment and affecting birds and their reproduction (by thinning eggshells), notably the bald eagle (Carson 1962). Arguably, Carson's careful research led to many international developments, including the introduction of the U.S. Endangered Species Act and the formation of the U.S. Environmental Protection Agency (EPA).

Much has changed since the 1960s. In 1967, the bald eagle was one of the first endangered species to be listed on the Endangered Species List, and on June 30, 2007, it was removed from the list. In the United States, DDT was among a dozen "persistent organic pollutants," or POPs, banned between the 1970s and mid-1980s (Stockholm Convention 2001). The U.S. EPA recently reviewed and reregistered all approved pesticides as required by the Food Quality Protection Act of 1996, conducting risk assessments that consider both human and environmental impacts (FQPA 1996). Today's pesticides are less toxic and much less persistent. After they are applied to crops, most biocidal chemicals used today break down quickly into less toxic compounds, and less of their residue ends up on fruits and vegetables. More and more pesticides are designed to target specific pests by interfering with their biochemistry. They are designed to act only on those species that affect crops and thus are less harmful to people.

Ironically, the World Health Organization (WHO) recommends indoor residential spraying of DDT for control of mosquitoes that carry malaria in Africa. When used in the 1950s and 1960s, DDT successfully eradicated malaria in many parts of the world. One of the impacts of banning DDT included the spread of malaria in parts of the world such as Africa and India, where the prevalence of this devastating disease today can be as high as 50% or more. WHO promotes the use of DDT for malaria because it feels that the evidence shows that the benefits outweigh the risks. Weighing the risks of malaria against those of DDT led WHO to advocate its use under well-managed conditions, where DDT poses no harm to wildlife or humans (WHO 2006). From this point of view, the alternative to DDT spraying is widespread malaria outbreaks and millions of people dying from a preventable and treatable disease spread by mosquitoes. WHO feels that the potential cancer risks associated with exposure to DDT are low, and these must be balanced against the millions of people who would suffer and die if malaria-spreading mosquitoes were not eradicated.

Looking back at the example of DDT as a case study in chemical management that led to current environmental management generally, and for pesticides specifically, shows that much is learned from looking at the evidence and weighing the risks and benefits. Scientists have been studying the effects of DDT since the 1950s. While much is known about its effects on people and animals, uncertainty remains regarding specific effects, for example, the association of DDT with cancer. Several studies suggest that DDT exposure does not increase the risk of cancer, but a few studies indicate that it does (JMPR 2000). The concern about DDT, and its thinning of bird eggshells, motivated decades of research on pesticide behavior effects in mammals, people, the environment, drinking water, and foods. On the one hand, the risks associated with pesticides need to be managed, and they are managed by current legislation. On the other hand, some would argue that the risks of limiting pesticide use have implications for public health, not only in the case of DDT and malaria, but for farmers, farm workers, and their families who may use greater quantities of less effective substances because of their regulatory status (Gray and Graham 1995). DDT may not be highly toxic and, in comparison to death from malaria, demonstrates that the benefits of using substances sometimes outweigh the risks—in this case, weighing young children dying versus low-level cancer risks. The larger issue in the 1960s may have been the indiscriminate use of DDT, not its use for mosquito control. It is easier to see in hindsight, of course, than to predict the future. There are many variables to consider. One important lesson from looking at DDT is that, after fifty years of study, there is still uncertainty about the associated health effects for humans.

A key message from this discussion is that we should not let our concerns over toxicity overshadow consideration of important societal benefits associated with new technologies. Of course we should strive as a society to have safer products, but there are other benefits to consider, for example in energy technologies that reduce our carbon footprint or, in the case of DDT, save lives.

Society can benefit from new technologies, but we must ensure that we do not replace our existing problems with new ones that we do not understand and cannot manage. Nanotechnology development presents an opportunity to transform our society to a more sustainable technological future, but this requires a groundswell of activity to steer in this direction; otherwise, it will not happen.

Incorporating nanomaterials into products creates the opportunity to reduce health risk, reduce pollution, and save energy and scarce resources. By focusing our attention on these end points during the development process, we can realize this opportunity.

1.7 What Is Risk?

> If we understood quantitatively the causal relationships between spe-
> cific technological developments and societal values, both positive
> and negative, we might deliberately guide and regulate technological
> developments so as to achieve maximum social benefit at minimum
> social cost. (Starr 1969)

Much of this book focuses on risk analysis and its potential applications
for nanotechnology and nanomaterials. But first—what is risk? Defining risk
is not as straightforward as one might expect. Understanding risk is a complex, multidisciplinary endeavor. There are many dimensions: technical,
economic, social, and political, and these dimensions are not universal and
are often divergent. This book adopts the view that while risks are shaped
by the societal and political context in which they occur, there are discern-
able physical and biological impacts associated with exposure to substances
that can be defined and assessed. This view is born from years of working in
the field of risk analysis seeking practical approaches to assessing risks associated with emerging potential threats in the environment as a first, essential
step. Thus, as may already be apparent, the complex societal dimensions of
risk are often noted, but not broadly considered as part of this discussion.

Early definitions of risk simply focused on the number of deaths associated with a particular hazard (Starr 1969). Over the years, the understanding of risk has evolved to encompass a much broader and more precisely
defined range of meanings. Risk may be defined differently, depending
on the context. For example, the U.S. Nuclear Regulatory Commission
defines risk as "the combined answers to 1) What can go wrong? 2) How
likely is it? and 3) What are the consequences?" (NRC 2007). The U.S. EPA
defines risk in the context of human health as: "The probability of adverse
effects resulting from exposure to an environmental agent or mixture of
agents" (U.S.EPA 2007). WHO currently defines risk as "the probability of
an adverse effect in an organism, system, or (sub)population caused under
specified circumstances by exposure to an agent" (WHO 2004).

At Clark University in 1985, the founders of one of the earliest Science
Technology and Society programs, Chris Hohenemser and Bob Kates,
with others, broadened the scope of understanding risk beyond its early
definitions. They defined risk as the "quantitative measure of hazard consequences expressed as conditional probabilities of experiencing harm"
(Kates, Hohenemser, and Kasperson 1985). Social scientist experts in risk
perception adhere to a broader definition of risk. According to Paul Slovic
(2000), Ortwin Renn, and other experts, risk is a construct (IRGC 2006).
That is, risks are judged in the context of individual and cultural views
of the world. The International Risk Governance Council (IRGC), based

in Switzerland, defines risk as "an uncertain consequence of an event or an activity with respect to something that humans value (definition originally in Kates et al. 1985). Such consequences can be positive or negative, depending on the values that people associate with them." This brings out a highly significant point: Risk has a societal dimension, a context for individuals and for groups with specific points of view. As a construct, there are many dimensions of risk that are not universal; rather, these are personal, developed in response to a number of factors, including whether people feel they have control over a hazard or feel it is imposed on them, and how scary a hazard is perceived to be based on an individual's level of experience with it. An example is traveling by air versus on the road. Statistically, fatalities are higher per mile driven on a road than flying in an airplane. But many people are more concerned about their safety on airplanes than when driving because of the familiarity and feeling of control over the potential risk associated with driving versus flying.

People's perceptions of risk are also influenced by their peers and by the media. Concern levels about risks decrease when the risks are more familiar, especially if they are associated with valuable benefits. For example, exposure to radiation as a cancer treatment is a more acceptable hazard than exposure to radiation from spent nuclear fuel. Societies continually manage many hazards, some better than others (e.g., food safety, water quality, terrorism, and air traffic). As a result, societies have defined acceptable levels of risk for many substances and technologies that people are willing to bear (Kates et al. 1985). Because of the role of perception though, the levels of acceptability can vary for different concerns. In this context, it is clear that early views of risk analysis simplified the societal dimensions.

1.8 Risk Analysis

Having introduced the concept of risk, now we can consider risk analysis and its role in nanotechnology and nanomaterials. Risk analysis is a multidisciplinary approach to understanding how substances behave and to judge whether that behavior is acceptable. Risk analysis involves both science and judgment, and this is part of the reason for its controversial nature. The science of characterizing materials, their toxicity, and their exposure characteristics is weighed against other materials and standards established as acceptable in a society. In the view of the WHO, risk analysis is "a process for controlling situations where an organism, system, or (sub)population could be exposed to a hazard. The risk analysis process consists of three components: risk assessment, risk management, and risk communication" (WHO 2004). Risk assessment is thus a key part of risk analysis.

This book explores risk analysis and risk assessment in much greater depth in subsequent chapters. Suffice it to say, risk analysis is a way of evaluating and weighing benefits and environmental concerns in a consistent and transparent way. It helps to balance perception with scientific analysis, and to consider substances and technologies through a framework that allows clearer decision making about their potential to cause harm to health and the environment. In the chapters that follow, the premise and mechanics of risk analysis are developed, along with examples of past, current, and future technologies that demonstrate the need for, and benefits of, evaluating the health and environmental risks of nanotechnology.

1.9 Overview of the Book

This book discusses nanotechnology and risk analysis, and, how through their marriage, a sustainable future can be built by design. We consider why we *must* proceed this way, the risks of not addressing health and environmental concerns, and ideas on how to go forward from here.

Chapter 2 explores the use of risk analysis in decision making and its development as a field of analysis and a policy tool. The chapter describes the steps of risk analysis, what types of information are developed and used, and how uncertainty is addressed. Chapter 3 looks at the "opportunity costs" inherent in current nanotechnology development, and explores in depth the possibilities for using nanotechnology to create a sustainable future. Chapter 3 also discusses life cycle analysis, an assessment approach that takes a broader look at the behavior of substances from their generation to ultimate disposal or reuse. Chapter 4, contributed by Dr. Richard Pleus of Intertox, introduces the topic of toxicology of nanoscale materials, what is known about the impacts of specific nanoscale materials on people, and key questions for further research.

Chapter 5 addresses environmental impacts and exposure—a crucial component that distinguishes *hazard* analysis from *risk* analysis. Exposure assessment looks in detail at the behavior of substances in the environment, including in occupational and ambient systems. There is somewhat of a convergence of thinking across sectors about evaluating nanomaterials and nanotechnologies throughout the life cycle, from manufacture to disposal. Chapters 6 and 7 explore tools adapting life cycle thinking into risk analysis for nanotechnology. These approaches represent the state of the art for assessing the risks of nanotechnology, but also require corollary risk management responses. Chapter 6 introduces NANO LCRA, the author's proposed framework for nanotechnology that incorporates adaptive management and life cycle thinking into a streamlined screening-level risk assessment process. In Chapter 7, Dr. J. Michael Davis describes alternative methods for evaluating risks of nanoscale materials and nanotechnologies, including Comprehensive Environmental Assessment. Dr. Thomas Peters of the University of Iowa contributes Chapter 8, which describes current practices for managing the hazards and risks of nanoscale materials in the occupational environment—who is doing what in this arena, and the state of the art. Chapter 9, coauthored by Professors Ann Bostrom and Ragnar Löfstedt, discusses risk communication and how people perceive the risks of nanotechnology. Finally, in Chapter 10, we survey the current state of numerous efforts internationally to address risks and develop science and policy for nanotechnology.

References

ASTM. 2007. Standard terminology relating to nanotechnology. International Standard E2456-06.

Auffan, M., Rose, J., Bottero, J. Y., Lowry, G. V., Jolivet, J. P., Wiesner, M. R. 2009. Toward a definition of inorganic nanoparticles from an environmental, health and safety perspective. *Nat Nanotechnol.* Oct; 4(10): 634-41.

Boyd, A. M., D. Lyon, V. Velasquez, C. M. Sayes, J. Fortner, and V. Colvin. 2005. Photocatalytic degradation of organic contaminants by water-soluble nanocrystalline C_{60}. American Chemical Society meeting abstracts. San Diego, CA.

Carson, R. 1962. *Silent spring.* New York: Houghton Mifflin.

Colvin, V. 2006. *Nanomaterials in the environment: An application and comment on implications.* Nano and the Environment Workshop, Brussels, Belgium.

Creighton, M. 2002. *Prey.* New York: Harper Collins.

Drexler, D. 1986. *Engines of creation.* Oxford, England: Oxford Press.

Feynman, R. P. 1959. Plenty of room at the bottom. Speech to the American Physical Society. http://www.its.caltech.edu/~feynman/plenty.html

FQPA. 1996. Pesticide legislation: Food Quality Protection Act of 1996. CRS Report No. 96-759 ENR 1996-09-11. Washington, DC.

Gray, G. M., and J. D. Graham. 1995. Regulating pesticides. In *Risk vs. risk: Tradeoffs in protecting health and the environment,* ed. J. D. Graham and J. B. Wiener, Chap. 9. Cambridge, MA: Harvard University Press.

Hullman, A. 2006. The economic development of nanotechnology: An indicators based analysis. European Commission Report. http://cordis.europa.eu/nanotechnology

IRGC. 2006. White paper on risk governance: Towards an integrative approach. http://www.irgc.org/IMG/pdf/IRGC_WP_No_1_Risk_Governance__reprinted_version_.pdf

ISO. 2010. TC 229. Core Terminology. ISO/TS 80004-1:2010.

JMPR. 2000. Joint Meeting on Pesticide Residues. Pesticide residues in food: DDT. http://www.inchem.org/documents/jmpr/jmpmono/v00pr03.htm

Kates, R. W., C. Hohenemser, and J. X. Kasperson. 1985. *Perilous progress managing the hazards of technology*. Boulder, CO: Westview Press.

NNI. 2012. What is nanotechnology? National Nanotechnology Initiative. http://www.nano.gov/nanotech-101/what/definition

NRC. 2012. Risk assessment defined. http://www.nrc.gov/reading-rm/basic-ref/glossary/risk.html

Roco, M. 2004. Interview with the National Nanotechnology Initiative. http://www.purdue.edu/discoverypark/dls/NNI/mikeroco.html

Roco, M., C. Mirkin, and M. C. Hersam. 2011. Nanotechnology research directions for societal needs in 2020: Summary of international study. *J. Nanopart. Res.* 13:427–45.

Slovic, P. 2000. Rational actors and rational fools: The influence of affect on judgment and decision making. *Roger Williams University Law Review* 65(1): 163–212.

Starr, C. 1969. Social benefit versus technological risk. *Science* 165(3899): 1232–38.

Stockholm Convention on Persistent Organic Pollutants (POPs). 2001. http://www.pops.int/documents/convtext/convtext_en.pdf

U.S. EPA. 2007. Integrated risk information system. Glossary of terms. http://epa.gov/risk/glossary.htm

U.S. EPA Office of Pesticide Programs. 2011. Regulating Pesticides that Use Nanotechnology http://www.epa.gov/pesticides/regulating/nanotechnology.html

Utsunomiya, S., K. A. Jensen, G. J. Keeler, and R. C. Ewing. 2002. Nanoscale uranium and fulleroids in atmospheric particulates. *Environ. Sci. Technol.* 36(23): 4943–47.

WHO. 2004. International program for chemical safety: IPCS risk assessment terminology. Part 1: IPCS/OECD key generic terms used in chemical hazard/risk assessment. World Health Organization. http://www.who.int/ipcs/methods/harmonization/areas/ipcsterminologyparts1and2.pdf

WHO. 2006. WHO gives indoor use of DDT a clean bill of health for controlling malaria. World Health Organization. http://www.who.int/mediacentre/news/releases/2006/pr50/en/

WWCS. 2012. A nanotechnology consumer products inventory. Project on Emerging Nanotechnologies. http://www.nanotechproject.org/inventories/consumer/

Further Reading

Frazier, L. 2001. Titanium dioxide: Environmental white knight? *Environ. Health Perspect.* 109 (4): A174–77.

Löfstedt, R. E. 2005. *Risk management in post-trust societies*. New York: Palgrave Macmillan.

2

Defining Risk Assessment and How It Is Used for Environmental Protection and Its Potential Role for Managing Nanotechnology Risks

Jo Anne Shatkin

CONTENTS

We should be guided by the probability and extent of harm, not by its mere possibility.

Aaron Wildavsky

There are two main dimensions of risk—the probability of an event occurring and the magnitude of the consequences. However, as discussed previously, the analysis of risk also includes judgments about the severity of risk as part of the assessment. This chapter walks through the basics of risk analysis, how risks are analyzed, and how scientists and regulators make decisions about how to manage them. Adopting the World Health Organization (WHO) definition, the risk analysis process consists of three components: "risk assessment, risk management, and risk communication" (WHO 2004). Risk assessment is a key feature of risk analysis. Several alternative frameworks are introduced for assessing and managing risks from substances and technologies.

We can never completely eliminate hazards. One can never guarantee a completely safe existence; the risk will never be zero. Even if you were agoraphobic and never left the house, you could choke on a sandwich, slip and fall down the stairs, get shocked by a loose wire. A logical way to minimize risk is to manage and prevent exposures to hazards. Risk assessment is a tool for analyzing hazards and the potential for exposure that informs such decision making.

One of the key features of risk assessment is that while there is always potential for hazards to occur, there is a difference between hazard and risk, and that difference is *exposure*. If there is no exposure to a hazard, then there is no risk from it. The level of risk associated with a hazard is dependent on the extent, or *likelihood*, of exposure. Explicitly considering exposure is the main difference between assessing hazards and assessing risks. For example, people with infectious diseases are encouraged to stay home from work. By showing up, they would expose coworkers to the disease, increasing the risk that they too would contract it. Without exposure, the coworkers are not at risk of getting infected.

Here is an example. The media has publicized studies showing that chopping (or cutting) boards can retain bacteria from foods prepared on them. Some of the bacteria, such as those in raw meat, can cause gastrointestinal illness. Disease-causing bacteria are a hazard. They could contaminate other foods prepared on the cutting board that are then eaten. However, in order for there to be exposure to the bacteria, the board must not be cleaned in between use. Usually, soap and water are adequate to remove the bacteria. So, if you wash your cutting board in between using it to prepare raw meat and other foods, your exposure to the bacteria is prevented, or reduced, and your risk of getting food poisoning is low. The ways in which different hazards are managed depends in part on where they occur (in the environment and in the world) and what kinds of hazards are being managed. There are many inconsistencies in society's level of concern, and the safety standards applied, for different hazards. Governmental programs and even individual actions can be inconsistent about the level of protection from different hazards. Some regulations require risks to be "as low as reasonably achievable," for example, while others set standards based on a quantitative measure of risk.

Some of the inconsistency is determined by regulation. If you live in California, the standard for exposure to products containing cancer-causing chemicals—one in a million at risk—is determined to be an acceptable risk level (the scientific notation for this is 10^{-6}). Note that this does not mean one person in a population of one million people *will* get cancer, it means one person of 1 million exposed is *at risk* of getting cancer from exposure. In the same way that being exposed to someone with an infectious disease does not mean with certainty that you would be infected, being exposed to a carcinogen does not mean you will get cancer. An exposure increases the risk of getting cancer.

In occupational environments in the United States, OSHA (the federal government's Occupational Safety and Health Administration) generally regulates carcinogens in the workplace to a risk level of 1 in 1,000 (10^{-3}), which is a 1,000 times higher risk than environmental standards in California. The U.S. Environmental Protection Agency (U.S.EPA) sets cleanup levels for hazardous waste sites to cancer risk levels between 1 in 1 million (10^{-6}) and 1 in 10,000 (10^{-4}). A 10^{-4} risk level might also be expressed as a 99.99% chance of not getting cancer from the exposure (10^{-6} is a level of protection of 99.9999%). These risk levels are on par with the risk of dying in an automobile crash in the United States (10^{-4} in the year 2000, before everyone talked on cell phones), a bicycle crash (10^{-4}), being struck by lightning (10^{-5}), or winning the lottery. Contrast these (low) risk levels with the lifetime risk of getting cancer in the United States of 1 in 3.5, which is less than 1% per year (10^{-2}), or the annual incidence of malaria worldwide (less than 1 in 10, under 10% or 10^{-1}). In parts of Zambia, the risk of contracting malaria exceeds 100% for children under five, meaning many may be infected more than once. The probability of infection in some provinces is observable, a far cry from the theoretically calculated extrapolations to the risks from low levels of environmental contaminants (Finkel 2007).

Clearly, these threats are not evenly distributed across time, space, or the type of hazard. However, one commonality from each of these hazards to health is that exposure is required to be at risk. There are hazards in daily life, and these differ depending on specific life circumstances. But, if you are not exposed to a hazard, then you are not at risk from it.

It is a key point, so let us reiterate: There are hazards everywhere, all the time. Risk is distinct from hazard because risk also considers the potential for *exposure*. Hazards alone do not constitute risk: There must be both hazard and exposure potential for there to be a health risk. That is why many of our current products and processes use very toxic materials without harm to people or the environment, because there is no exposure. The risk assessment paradigm includes the two dimensions of hazard and exposure. Risk analysis offers more opportunities to manage hazards safely. One option is to eliminate the hazard, but another is preventing exposure to it.

Some argue that it is preferable to remove hazards rather than prevent exposure to them. This approach has appeal, and I agree that ultimately it is safer to not have hazardous materials in products. However, looking at hazard and exposure can tell a different story than one judged on the basis of hazard alone. Adopting a broader view of potential impacts, considering both hazard and exposure, can lead to more informed decisions about the potential impacts of choosing one material, or technology, over another. Below is a brief discussion of risk perception and its role in risk analysis. A more in-depth discussion is presented in Chapter 9, contributed by Professors Ann Bostrom and Ragnar Löfstedt.

2.1 Context for Technological Risk

A 2006 survey to assess people's perceptions of the risks of a range of technologies found that U.S. respondents thought nanotechnology was scarier than bicycles (Nanotechweb 2006; Curall et al. 2006). What is the relationship between bicycles and nanotechnology? Both nano and bicycles may be considered enabling technologies; that is, their use enables other applications. A bicycle is a technology, but it is not a new one. Even though components, frames, and accessories are often made with state-of-the-art materials, bicycles are part of a way of life in societies around the globe. Certainly, while nanotechnology is used for stronger, more flexible, and lighter frames for bicycles, cycling is not new, bicycles are not new, and to many, they are simple machines that create an effective means of transportation.

The modern bicycle was invented in the late nineteenth century. While continuously evolving in the materials, components, and aerodynamic design, bicycles are common enough that each of us likely conjures up image that includes two wheels, a frame, seat, and handlebars. Some applications of the bicycle are for transportation, while others are for exercise, and for others, professional employment (messengers, racers). When someone says "bicycling," it is easy to conjure up an image of what they are describing. These might be leisurely rides along canals in the Netherlands, crowded city streets in China, the hilly back roads in New England, the sleek bodies of racers climbing the mountainous roads in the Alps. What does the word *bicycling* conjure up in your mind?

Now, let us try nanotechnology. Does the word *nanotechnology* conjure up a specific image for you? Do you think of coated fabrics, tiny sensors,

super-light aircraft, golf clubs, bacteriostatic medical devices like cathe-
ters, or sunscreen? Do you think of molecular machines? Bacteria in a fuel
cell churning out energy? Does nanotechnology conjure up spybots, as in
Prey (Creighton 2002)? Nanotechnology involves creating molecules that
enable applications in medicine, energy, coatings, sensors, electronics, fabrics, polymers, and industrial uses. Although diverse in terms of composition, structure, and physical properties, nanotechnologies can be defined
by their size and through their applications. But it is not as easy to conjure
up an image of "nanotechnology" as it is to envision a bicycle. One can
have a different image every time a report of a new application appears.

These represent images of the technologies, but what about the risks?
The 2006 survey compared people's perceptions of nanotechnology to a
host of others and found people to be fairly neutral about nanotechnology risks. In fact, survey respondents were less concerned about nanotechnology than the insecticide DDT (Curall et al. 2006). The perception
of technological risk has a lot of influence on individual and collective
decisions about whether or not to adopt new technology. Perceptions of
technological risk can also influence how extensively risk assessments are
conducted. The assessment approaches used also influence the decisions
about management approaches adopted to make them safer. The level of
threat to something valued (e.g., health, loved ones) is a key driver in an
individual's assessments of technological risk.

As we discuss later in this chapter, there may be differences between
the level of concern consumers and others have about a substance or a
technology, and the views of those with a lot of expertise in it, including
the developers. Those involved in regulating or managing risks from a
technology will be influenced by both public opinion and industry views.
Often, the level of public concern is a key determinant in how extensively
the risks from one technology versus another are managed. Like it or not,
public concern is a driver of governmental efforts in a democratic society.
That is, perception and public opinion can cause us as a society to focus
attention on the wrong risks, ones that are not as important to protecting
public health. For example, when Dr. Oz tells people that they should not
give children apple juice because it contains arsenic, people might switch
to soda, which has no health benefit and is linked to diabetes and obesity.
The truth is that there is no health risk from arsenic in apple juice, because
it is not toxic in the form that occurs in apples. Irresponsible information
about health risk shifts priorities away from important concerns to popular and sensationalized ones. This is important to our discussion, because
focusing on risks that do not really affect public health, safety, and the
environment can come at the expense of those that do.

The current U.S.EPA regulatory process is driven by the concerns of the
American public that prevailed in the 1970s and 1980s. The Comprehensive
Environmental Response, Compensation, and Liability Act (CERCLA) or

Superfund program, for example, requires that hazardous waste sites be remediated when they pose risks above an acceptable range of 1 in 1 million (10^{-6}) risk to 1 in 10,000 (10^{-4}) risk of getting cancer from exposure to contaminants in soils and/or groundwater (in some cases also from rivers and lakes and from air pollution) on the site for a lifetime. Compared to the risk of harm from riding a bicycle, those are pretty long odds. Over 1 million people *in a year* are injured on bicycles (Petty 1991). Yet many people fear the effects of hazardous waste, and the U.S.EPA oversees the cleanup to ensure that Superfund sites pose residual risks that are extremely low. From a utilitarian view, to protect people's health it would make more sense to invest in bicycle safety than cleanups of low-risk hazardous-waste sites. This is not to suggest that cleaning up pollution is not an important priority—only that using risk assessment helps to align resources with health and safety impacts. Understanding the role of public concern in the assessment of risk is a key dimension.

Returning back to the comparison of risks from bicycles versus nanotechnology, the Centers for Disease Control estimates that 67 million Americans ride bicycles. In the year 2005, 784 people were reported to have died in bicycle crashes. Therefore, the risk of dying in a bicycle crash is about 1 in 10,000 (10^{-4}). Are people afraid of dying on a bicycle? Apparently not, since only about 19% of adults and 13% of children who ride bicycles wear helmets consistently (CDC 1999). How many people are dying from nanotechnology in a year? Currently, none. Is it likely that hundreds of people each year will die from nanotechnology? If not, then what causes people to fear nanotechnology? And, why are we not more afraid of bicycles?

Is it fair to compare the risks from nanotechnology to bicycles? The risks of bicycles are very well known, as are the benefits of bicycle riding. The risks and the benefits of nanotechnology are yet to be determined. It is fairly obvious that people generally have a lower level of concern about familiar technologies versus ones not well known with little experience. In a survey of sixteen technologies, Morgan et al. (1985) found bicycles were among the lowest of the dreaded risks and also among the most known. When the risks from bicycles are among the least dreaded and the best known, how surprising is it to learn that nanotechnology risks are perceived as greater? The finding does not suggest concern about nanotechnology as much as it indicates a lack of familiarity. It does suggest that as people learn more about nanotechnology, perceptions may change.

As you continue reading and form your own opinions about nanotechnology risks, here is a point to keep in mind. The modern bicycle became popular in the late 1890s, including among women. The famous suffragist, Susan B. Anthony, said, "Let me tell you what I think of bicycling. I think it has done more to emancipate women than anything else in the world. It gives women a feeling of freedom and self-reliance."

In the nineteenth century, few could envision that a technology like the bicycle could change American culture (Anon. 2007). Women gave up corsets and long gowns in favor of more practical clothing so they could ride bicycles. My point is, as with risks, the benefits of new technology are also uncertain and may not be predictable. When there is little knowledge, it is important to keep an open mind, and readjust thinking as new information is obtained, to be adaptive in our thinking.

2.2 Why Risk Assessment for Nanotechnology?

Each of us conducts risk assessments based on our own judgments. A successful venture capitalist told me once that he has two very simple decision tools regarding whether a new nano-product he may invest in has the potential for health and environmental risk: (1) "Is the product free in the air, are there free particles?" and (2) his gut reaction to the question, "Is it too risky?" This is indeed conducting risk assessment. However, without formal tools, he is unable to document the decision process, and perhaps has less confidence in his decision than if a formal analysis had been conducted that considers not only the available data, which are sparse, but the characteristics of the material. In answering the question about free-particle exposure, one type of risk assessment is conducted: Will the material present an inhalation hazard in the workplace? This is an important question, but as explored in later chapters, determining that a nanomaterial does not pose an inhalation hazard is not the same as concluding a lack of health or environmental risk. Other pathways, as well as the potential for exposure outside the research and development and manufacturing environment, are also key factors that must be addressed.

As discussed in Chapters 6 and 7, an alternative is to adopt a new risk analysis process for nanoscale materials; to conduct a more comprehensive assessment of their potential for harm to health, safety, and the environment; and to make the assessment iterative, improving analysis as more data become available. The current regulatory framework for substances and technologies is a patchwork of levels of protection. Substances in drinking water are allowed at much higher risk levels than in soils on a hazardous waste site. Occupational risk standards can be much less protective than permitted releases to the environment, where exposures are more diffuse. To address these inequities, risk assessment is increasingly the basis of standard setting.

Risk assessment helps to identify potential concerns and evaluate how they compare with other types of materials and technologies that have been adopted. Using screening tools that evaluate risks across the product life cycle during product development helps in making sound decisions about adoption. This achieves a better level of protection for people and the environment, with a rational basis to compare one risk to another, and an even playing field for newcomers compared to existing substances and technologies.

One benefit of using risk assessment to set standards is that it allows decision making under uncertainty. That is, often decisions need to be made before all information is known. If the details were known, the discussion would not be about risks; it would be about safety. But there is not always time or other information resources to allow complete understanding. Risk analysts have dealt with this problem by making reasonably conservative assumptions that tend to overestimate risk, but yet allow decisions to be made in the absence of a complete database of information.

Nanotechnologies are not specifically regulated today, although there are many developing regulations for reporting and testing of nanomaterials in California, Europe, Canada, the United States, and elsewhere (see Chapter 10). One reason for the lack of regulation is that it is so early in the development and use of nanomaterials in technology that there is not enough experience to discern which aspects of nanomaterials need to be differently regulated, or how best to regulate them. As discussed in Chapter 4, many studies have measured the effects of nanoparticle exposure on health, but they often suffer from methodological concerns and, in many cases, raise more questions than they answer. Governmental authorities have confirmed that general safety laws apply to nanotechnologies and nanomaterials, but developing new and specific rules takes time and is subject to deep scrutiny and objections from industry and advocates. In this uncertain environment with dynamic regulatory requirements, risk assessment provides a transparent process to frame and characterize risk, and can inform sound policy development.

Throughout the book we examine the consequences of not identifying potential problems early. Nanotechnology eventually may affect everyone. Instead of pretending there will not be any problems, it is time to look, to begin to identify concerns early, in time to take steps to address them. Unless we are looking, we will not know whether the products are safe, even if the necessary tools to see all of the details are not yet available.

Risk assessment allows us to prioritize the gathering of information needed for sound decision making. Stepping through the assessment process identifies where the missing pieces of information are and how important they are to overall decision making. If the key question relates to what happens to a material when it is released in water, then experiments can be designed to ask and answer this question. Before beginning lengthy toxicology studies, it is important to first consider the real-world conditions for nanoscale substances and nanotechnology. How will these materials be used? By whom? How much contact would there be? Characterizing the potential for exposure is necessary in terms of conducting good toxicology studies, and potential exposure can be identified easily in screening-level risk assessment frameworks.

One reason for conducting early assessments of risk is to distinguish perception from reality. As discussed previously, what people worry about is an artifact of who they are, their social network, and where they get their information. People's perception of what is harmful may not match the statistics in terms of probability of harm. Risk assessment provides a sound basis to clarify what is harmful and what is not, even when there are data gaps. Choosing to use risk assessment yields many benefits, including early identification and prioritization of health, safety, and environmental concerns.

Risk assessment allows examination of the balance of risk/benefit trade-offs. This does not necessarily translate to direct economic benefits but, rather, emphasizes that the introduction of new technology means replacing existing technologies, and there can be health and environmental benefits of doing so. A new substance that can reduce dependence on a very toxic chemical reduces risk; even if the new substance is not completely benign, there is still a benefit.

There are other reasons to conduct risk assessments, even if they are not specifically required for the approval of a nanoscale material or technology. One is to demonstrate commitment to regulators and the surrounding community. Even if a producer is not required to look at the risks from products, it might make sense to do so in order to inform stakeholders—people who are concerned about or responsible for effects from products—of the state of knowledge. If you own a manufacturing plant, neighbors could be concerned about what is released into the air and water, what is stored on-site, and what would happen in the case of a fire or explosion. Conducting a risk assessment allows you to communicate about these concerns and promote healthy relationships with the community.

Another benefit of risk assessment is to allow comparison of alternative management strategies. There is always more than one way to solve a problem, and it is best to be informed by the available data when making decisions. A technology may require working with a material of unknown toxicity, and how best to handle it can be informed by examining the potential for exposure and risk. Knowing what the concerns are creates an opportunity to address them in a proactive manner. Even if the available data do not allow quantitation, risk assessment informs effective risk management, and addressing risks earlier is cost efficient and responsible.

2.2.1 Adaptive Risk Assessment for Nanomaterials

Here is where I believe risk assessment makes sense: to inform the decisions about how to manage nanomaterials and nanotechnology amid uncertainty. As you might imagine, it is not a simple question of whether nanotechnologies are safe or toxic; there is a whole spectrum of more likely possibilities in between these two extremes. Even with significant uncertainty, a risk-informed evaluation makes sense.

There are many ways to conduct risk assessments. One important first step, especially for nanotechnologies and nanoscale materials, is to conduct a *screening-level assessment* (described in Chapter 6). The ease of modifying nanoscale materials through engineering makes it overwhelming to consider detailed quantitative risk assessments for every type of material, prior to commercial introduction. However, it would be very useful to conduct screening-level risk assessments for new nanoscale materials, since this would allow the assessments to keep pace with the rapid developments in nanotechnology.

The real value of conducting early screening-level risk assessments is that understanding risks allows more efficient management of them. Looking for potential problems early reduces the potential for unforeseen impacts. Customers, regulators, manufacturers, and activists will have increased confidence about the safety of new products if the concerns about them are assessed and addressed early. It is also important to *consider the entire life cycle* of a material to understand the potential for impacts to the environment. What is the life cycle? Some refer to it as "cradle-to-gate" or "cradle-to-cradle." Considering the potential for effects throughout the life cycle is an important step in generating assessments for new materials. Life cycle thinking and screening-level life cycle analysis is discussed in Chapter 3.

2.3 Origins and Development of Risk Assessment and the Societal Dimensions of Risk

For decades, engineers have been making calculations about the strength of support beams for buildings, stability of bridges, and the crash resistance of automobiles, among others. The U.S. Food and Drug Administration (FDA) reviews studies of the safety of drugs, food additives, and medical devices, among others. The Food and Agriculture Organization (FAO) devises international protocols for identifying and managing a range of threats to the food supply. Public health officials study the outbreak of diseases associated with various exposures. These efforts inherently involve the assessment of risk.

However, many people conduct risk assessments all the time. Will the stock market go up or down? Is it safe to cross the street? How long has that leftover been in the refrigerator, and will it make me sick if I eat it? Is it safe to drink the water from the tap? Will I need an umbrella today? Can children safely use these products? Are people adequately protected from exposure to these materials in the lab? Individually, each of us might answer some of these questions differently. You might be unconcerned about the safety of your drinking water, but your best friend may insist on bottled water for drinking (which may or may not actually be safer to drink because it is not regulated). Regardless of whether the assessments calculate probabilities, we judge the likelihood and the consequences of our actions and the actions of others, and use that judgment to make decisions. That, in a nutshell, is risk assessment.

Chauncey Starr, founder of the Electric Power Research Institute, is considered the godfather, or grandfather, of quantitative risk analysis. His seminal 1969 paper, published in the journal *Science*, described what is still recognized as the key principles of risk analysis. Starr describes a quantitative analysis of the probability of dying from an industrial accident, and relates it to the exposure and the length of time someone works. This approach formed the foundation for the current approaches to risk analysis.

Interestingly, Starr conducted this analysis to evaluate the social acceptance of risk—an integral, but less explicitly discussed concept that, as a society, people are willing to accept some level of risk when the benefits of the risk source are valued. Starr was indeed exploring the nature of the technology/society interface, yet the main result of his work was the use of quantitative measures of exposure and effects to estimate risk. The basic principles of risk are key concepts that are revisited later in this chapter. Starr noted that the perception of risk is different when the risks are voluntary or chosen versus those that are imposed. This, in part, explains why some people are afraid of low-probability risks, such as a nuclear power plant failure, but not of higher probability risks, such as developing lung cancer from smoking cigarettes (Starr 1969).

In 1983, the National Academy of Sciences convened experts to look at how risk assessments were being done by the U.S. government. The report, *Risk Assessment in the Federal Government: Managing the Process* (NRC 1983), is also commonly called the Red Book. A key theme of the Red Book was the distinction between the process of risk assessment and that of risk management. Risk assessment should be done by people different than those responsible for making decisions based on the assessments and those managing the risks. Thus, the assumptions and conclusions that constitute risk assessment must be independent of the broader management and policy. The reason is to keep the analysis independent of external concerns, such as economics or political pressure that can factor into decision making about risks. The Red Book laid out four main steps for risk assessment: hazard identification, dose-response assessment, exposure assessment, and risk characterization. This process of risk assessment

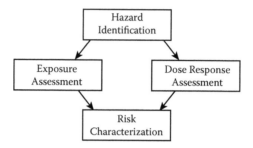

FIGURE 2.1
Basic four-step process of risk assessment.

has become the foundation of both voluntary and regulatory policies for analyzing risks and developing quantitative estimates of risk that inform decision making.

Figure 2.1 shows the basic four-step process of risk assessment. In the thirty years since the Red Book was published, much work has been done to refine the processes and approaches for these steps, which are described in more detail later in this chapter. Some frameworks use different terminology, and others outline steps for including stakeholders (people affected by decisions, or those with responsibility to manage or bear the results) or revising assumptions, but this basic model remains the current approach for using risk assessment for decision making. The National Research Council affirmed this in their evaluation of how agencies are using risk analysis (NRC 2009). In *Science and Decisions*, the expert panel made a slew of recommendations to better inform the risk assessment process by framing the problem more directly and to better the treatment of uncertainty (Shatkin, 2009), but overall embraced the 1983 framework as sound.

2.4 Frameworks Addressing the Social Dimensions of Risk

One early contribution to the issues of society and technological risk was *Perilous Progress: Managing the Hazards of Technology* (Kates, Hohenemser, and Kasperson 1985). *Perilous Progress* laid out an approach for assessing technology and addressing the societal components of risk, using causal chain analysis. *Causal chain analysis* focuses on how to assess and manage technological risks. It identifies the alternative risk management options in terms of mitigation measures at different steps in the technology development and implementation process. As shown in Figure 2.2, technologies are developed in response to human wants and needs. Some hazards

FIGURE 2.2
Technologies are developed in response to human wants and needs.

can be managed by choosing alternative technologies. In other words, there may be more than one technological solution to address a particular need. Often, however, a technology is developed, and no effort is made to address the attendant hazard occurring until some *initiating event* occurs. In this situation, risks are managed by mitigating the consequences of that event. Evaluating technologies using causal chain analysis allows the opportunity to compare alternative management strategies for technological risks.

Perilous Progress also develops an approach for evaluating technological hazards in a societal context. It includes the work of Paul Slovic, Baruch Fischhoff, and Sarah Lichtenstein (Slovic et al. 1985; Kates et al. 1985), which addressed the social factors that contribute to the perception of technological risks. In surveys of laypersons and experts, they found that people with a limited understanding of the technical aspects of risk perceived some hazards such as nuclear power to be riskier than experts did. The hazards that people associate with risk factors, such as dread and unknown risk, were perceived at a higher level of risk than those based on the probability and magnitude of those hazards. This means that while experts tend to judge risks by the statistical probability of adverse events happening, other people may judge risks not by how likely they are, but by the type of risk—for example, how much is known about a hazard and its impacts; how much it is dreaded; the nature of the consequences, including the severity of effect; and the potential for catastrophic effects.

The significant body of work on the societal dimensions of risk is beyond the scope of this chapter, but it is discussed in detail in Chapter 9. The topic is introduced to raise awareness for the discussion of using risk analysis for environmental decisions. Risk analysis only informs better environmental decisions if it addresses the key concerns in the assessment. Having experienced the development of sound, scientific assessments of environmental concerns to inform governmental decision making, and then seeing them disregarded because of public pressure (generally leading to overly conservative measures to eradicate risk), has humbled my view of the role of science in environmental decision making. Yet, I have participated in decision-making processes that involve educating people about risk and risk analysis, and have seen that process lead to more-informed decisions about managing environmental risks. People need to be informed about risks and risk analysis so that they can make educated decisions. Even so, the power of people to influence decision making should never be underestimated, and there is a great need to raise the level of awareness about science and technology issues so that debates can focus on sound evidence.

As described in Chapter 9, researchers in the field of risk perception continue to survey attitudes toward technological and societal risks and refine the understanding of the factors that contribute to public perception of risk. While it may seem irrational to some scientists and engineers, public reactions are predictable by the nature of the risk, in terms of whether hazards are reversible, dreaded, the level of media attention to them, and the association to other types of hazards. The often-cited concern that the fate of nanotechnology will be like genetically modified organisms (GMO) in food relates to the framing of the GMO debate by Greenpeace, who termed GMOs in food as "Frankenfood" (Asian Economic News 2001). There are signs that some groups may take a similar approach with nanotechnology. While the surveys to date on public attitudes toward nanotechnology reflect a low level of understanding, they also measure a positive perception of the benefits of nanotechnology. However, in the information age, this can change rapidly.

As Kahan stated, "Not much more is known about public perceptions of the risks of nanotechnology than is known about nanotechnology risks themselves" (Kahan et al. 2007). In their survey of 1,800 people in the United States, they found strong opinions about nanotechnology risks, despite more than half of survey respondents never having heard of nanotechnology. This was affirmed in a meta-study by Simons et al. (2009), who compared surveys in Germany, Australia, and the United States and found that even when people did not understand nanotechnology, they were generally positive about its societal benefits. Interestingly, while those with more knowledge were more favorable toward the benefits of nanotechnology over the risks, the authors concluded that the real finding is that people's opinions about nanotechnology align with their cultural values, not with knowledge, meaning that whether nanotechnology is likely to become as controversial as nuclear power did is yet to be determined. One challenge for nanotechnology development is the need to evaluate risks at the same time as they are being developed, in real time. In the information age, where a video can be viewed by millions of people within hours of posting on the Internet, social networking allows unstructured communications not limited by geography, and the flow of information, whether it is true or not, is nearly instantaneous.

The case of the recall of a product called "Magic Nano" is instructive. In 2006, the German government recalled Magic Nano©, a bathroom cleaning product, after eighty people were hospitalized with respiratory symptoms after using it (Weiss 2006). Within days, the ETC Group called for a moratorium on nanotechnology. A month later, the German government declared that the product did not contain nanotechnology, that it was a flaw in the manufacturing of the aerosol propellant that caused adverse reactions. Scientific and government responses cannot match the speed of electronic communication.

One theme revisited in this book is the considerable uncertainty associated with the impacts of nanotechnologies on health and the environment. When there are missing data, risk assessments apply professional judgment and other tools to extrapolate risks. The judgments reflect values. For example, defining the significance of the risk presented by a nanoparticle measured in air becomes a question of the acceptability of the risk, which of course varies across organizations and individuals, depending on the level of concern. Because of the importance of societal dimensions of risk assessment, several risk frameworks incorporate public participation.

The NRC also addressed the issue of uncertainty in the 1996 report (known informally as the Orange Book), *Understanding Risk: Informing Decisions in a Democratic Society* (NRC 1996). This effort addressed the nature of risk assessment as an analytic-deliberative process—that is, it involves both analysis of the problem and discussion to reach agreement among people about interpreting the analysis. The decision-making process needs to focus both on the technical issues and on improving understanding and participation. According to the NRC panel, "Appropriately structured deliberation contributes to sound analysis by adding knowledge and perspectives that improve understanding and contributes to the exact ability of risk characterization by addressing potentially sensitive procedural concerns." In *Understanding Risk*, the NRC lays out elements of an analytic-deliberative process: getting the science right; getting the right science; getting the right participation; getting the participation right; and developing an accurate, balanced, and informative synthesis. All of this is to say: Be clear what problem you are solving, and ensure that it is the one that people care about and that people agree with how you are doing the assessment, what data are used, and how they are interpreted.

The societal dimensions of risk were addressed by the 1997 U.S. Presidential Commission on Risk Assessment and Risk Management report, *Framework for Environmental Health Risk Management*. Figure 2.3 shows a schematic of the proposed framework for risk management that puts stakeholders in the middle of the decision process, engaging their participation at each step of the process. This risk management framework is intended to be broad, to address a range of types of hazards, and to implement an iterative process that revisits the problem and the risk management options.

The U.S. Presidential Commission framework recognized the role of uncertainty in risk assessment: "Risk assessors have to use a combination

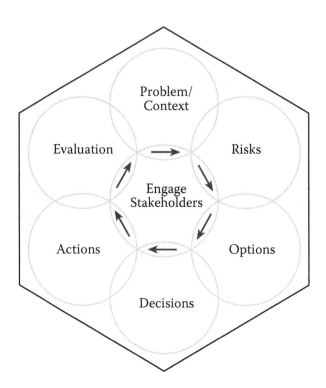

FIGURE 2.3
Proposed framework for risk management.

of scientific information and best judgment" (1997). Uncertainty is a key attribute of risk. If there was certainty about the impacts of a particular substance or technology, one would conduct a safety assessment and establish a definitive safe level. But with new materials, there is rarely that level of certainty that all of the relationships between exposure and effect are understood, and assumptions are made to address the inherent uncertainty. That is a main reason to involve stakeholders in decisions about managing risks. Stakeholder values and preferences must be considered in deciding how to manage risks under uncertainty.

The U.S.EPA acknowledges that environmental decision making can be controversial, involving not only science but also social and economic factors, political considerations, technological feasibility, and statutory requirements that may often be conflicting. The U.S.EPA "conducts risk assessments to provide the best possible scientific characterization of risks based on a rigorous analysis of available information and knowledge" (U.S.EPA 2004).

The International Risk Governance Council (Renn 2005) considered these and many other models in the development of their framework.

The main contribution of the IRGC framework is the inclusion of the societal context in risk assessment and risk management. In their governance framework, IRGC gives equal weight to the societal dimension of risk management, recognizing that some societal risks are more complex, and of greater concern, than others in a governance model. A major innovation of the IRGC framework is categorizing risk-related knowledge. Categorization addresses complexity, uncertainty, and ambiguity of risks. The IRGC framework also considers risk/risk and risk/benefit trade-offs. An example of a risk/risk trade-off is the risk of complications from surgery (the risks of complications may or may not outweigh the benefits of the surgery).

The IRGC has applied this framework to nanotechnology. In *Nanotechnology and Risk Governance*. Mikhail Roco and Ortwin Renn, pioneers in nanotechnology and societal dimensions of risk analysis, respectively, describe four generations of nanotechnology and their differences in terms of complexity, uncertainty, and ambiguity (Renn and Roco 2006). The first generation, passive nanostructures, represents those materials that exist or are in development today. The second generation involves active nanostructures, such as smart packaging or targeted drug delivery. The third generation (self-assembling structures) and fourth generation (molecular manufacturing) are viewed as forthcoming. Moving beyond the first generation of types of materials currently applied in nanotechnology (generally passive nanoscale particles, or substances and structures created at the nanoscale such as silver or gold that are smaller than larger particles, but pretty much remain as they were manufactured), the complexity, uncertainty, and ambiguity increase, and risk-governance models must adapt (Renn and Roco 2006).

The common themes in these various frameworks and approaches to risk assessment are: (1) the delineation of an analysis, generally the stepwise process relating hazards to exposure to assess risk, and (2) transparency in the process for making decisions about how to manage risks—including acknowledging the societal dimensions of risk—through a process that includes some level of participation by stakeholders.

2.5 How Risk Assessment Is Used in Environmental Decision Making

With some grounding in the broader societal dimensions, now the discussion hones in on how environmental risk assessment becomes a decision-making tool used to analyze and help make decisions about substances and technologies. Simply stated, risk assessment allows the estimation of health and environmental impacts from exposure to a substance.

Governmental and private organizations all over the world use risk assessment in environmental and public health decision making for risk management. In the United States, the U.S.EPA uses risk assessment to understand risks and make management decisions regarding cleaning up hazardous waste sites, closing municipal solid-waste landfills, setting standards and managing substances in drinking water, evaluating air quality, establishing tolerances for pesticides and other chemicals, and for setting policy on specific substances. Risk assessment is also used in the United States for regulating food safety and allowing additives in food and food packaging.

The European Commission passed chemicals legislation in December 2006 that requires registration of, and risk assessments for, all chemicals used in commerce in the European Union. The program, called REACH (Registration, Evaluation and Authorization of Chemicals), is viewed as a precautionary approach to chemical management and envisioned as a way to identify, prioritize, and manage chemical hazards. Depending on how much of the chemical is produced, different levels of testing and reporting are required. The assessments also depend on specific properties of chemicals, known either from testing or predicted by models (European Commission 2006). REACH is also an integrated framework that considers occupational exposure, consumer exposure, and environmental exposure. One key feature of REACH is that it includes a screening framework. Based on the results of the screening determination for a substance, a more detailed assessment is made (screening aspects of risk assessments are described later in Chapter 6) (European Commission 2006).

Risk assessments are also increasingly the basis for occupational exposure standards. In the United States, OSHA conducts risk assessments for substances to establish a standard for U.S. workplace environments. Occupational risk assessments typically involve studies of large groups of workers in specific industries, using epidemiology studies to relate workplace exposure to effects, as was done recently for hexavalent chromium. The U.S. National Institute for Occupational Safety and Health (NIOSH), a nonregulatory governmental organization, recently conducted a risk assessment for titanium dioxide in the workplace, specifically considering the evidence for nanoscale titanium dioxide and establishing a workplace exposure limit (NIOSH 2007), as discussed further in Chapter 8.

Risk analysis is also used for food safety. Decision-making governmental organizations deal with safety by conducting risk analyses of food ingredients and products. CODEX Alimentarius develops international standards under FAO and WHO, which are internationally adopted and generally focused on the biological safety of food (FAO 2005). The Organization for Economic Cooperation and Development (OECD) is an international organization that conducts many activities in its member countries to assess and evaluate chemicals, biocides, and other materials. OECD develops databases and tools and also harmonizes approaches to risk analysis (OECD 2007).

The American Society for Testing and Materials, ASTM International, has several committees that develop voluntary standards. One standard developed by committee E50 applies a tiered approach to risk analysis, called Risk-Based Corrective Action, for oil-impacted or hazardous waste sites. This approach applies a screening-level risk assessment first, which then leads to decision making and more detailed analysis where warranted (ASTM 2004). The U.S. FDA uses an approach called HACCP (Hazardous Analysis Critical Control Point) for food safety hazard analysis, combining assessment and management processes to determine the most effective way to manage risks in food safety (FDA 2001a, 2001b). Risk-based approaches are also used for managing medical devices and pharmaceuticals.

The U.S.EPA has published a process on how to conduct ecological risk assessments that has a somewhat different focus, because the assessments may evaluate risks to a population (e.g., a species of fish in a particular location) or an entire ecosystem made up of several levels of organisms within a food web. Ecological risk assessment has an additional layer of complexity over the assessment of risks to human health. It follows the same general process, but instead of defining a particular hazard, it begins by formulating a problem. The next step is to identify the assessment endpoints—in other words: What measures will be evaluated in the assessment? The measurement end points are the actual tests conducted to evaluate the end points that might include models and also field evaluations (U.S.EPA 1998).

2.6 The Four Steps of Health Risk Assessment

As we've discussed, if there were an absolute answer to what is safe and what is not, there would be no need to conduct an assessment of risk. However, where there is uncertainty in the information needed for making decisions, the information is interpreted, using judgments from prior experience. This uncertainty and individual attitudes toward risk affect how the data are judged and the conclusions reached about the significance of risk. It is the combination of scientific data and judgment about the data that constitutes the assessment of risk. Conducting formal risk assessments allows us to apply a process to determine the relative level of risk and, on that basis, judge whether the risk is acceptable or not.

As shown in Figure 2.1, there are four basic steps in a health or environmental risk assessment:

- Hazard identification
- Exposure assessment
- Dose-response assessment
- Risk characterization

First, the problem or the hazard is defined. The various international frameworks use different terminology for this step, referring to it as problem formulation, hazard characterization, or *hazard identification*. The step of hazard identification defines how to conduct the remainder of the assessment. It defines which questions the risk assessment will ask and answer. It provides a characterization of the hazard that describes its key attributes. First, define the nature of the hazard: Where is the contaminant? Is it in the air, in a product, or in the water? How much is present? Where is the contamination likely to go? What is the nature of the source material?

The second step is to develop an *exposure assessment*. Exposure assessment involves identifying the potential for an exposure to a hazard—a substance, a pathogen, or a technology—by considering all of the potential circumstances of exposure. Key steps include identifying receptors who may be impacted, pathways of exposure (where is the material released and how could one come in contact with it?), media of concern (soil, air, water, food, or product), and routes of exposure (can the hazard be inhaled, ingested, or absorbed through the skin?). Exposure assessment also considers the magnitude of potential exposure (how much exposure occurs), its likelihood of occurring, how frequently, and for what length of time. Exposure assessment can be very complex and often involves models and other estimation methods and assumptions.

In human health risk assessment, *receptors* are those who may be exposed to an agent under different scenarios, such as workers in an occupational environment who manufacture a chemical, users of a product, and others who may have incidental exposure occurring as a result of manufacturing, use, or disposal of a product. If a substance is in drinking water, the person who drinks the water is the receptor. In ecological risk assessment, the receptors may be specific species, populations, or entire ecosystems.

Substances have physical and chemical characteristics that govern behavior in the environment and will impact their behavior in the environment, so it is important to understand whether exposure is likely to occur by a particular pathway (e.g., inhaled, ingested, or absorbed through the skin). For example, a solid substance would have to somehow be released in the air that is breathed; otherwise, one would not evaluate an inhalation pathway for that substance. Generally, one of the key contributions to risk is the likelihood or extent of potential exposure, so defining the magnitude of exposure and how likely it is to occur is another very important step in characterizing risk. Even substances required in the diet, such as Vitamin A, can be toxic if overconsumed (this would amount to overexposure).

It is important to determine whether any exposure occurs at all. For example, if your neighbor uses an MP3 player with nanotechnology-enabled components inside his home, does this create an exposure for you? You would have to somehow come in contact with the nanotechnology parts in the MP3 electronic components, say by visiting your neighbor and destroying the MP3 player, to have an exposure. Again, without exposure, there is no risk. If, for example, your neighbor burns the MP3 player, then it might be possible to have exposure to a very small dose; otherwise, there would not be an exposure.

The next step in a risk assessment is to evaluate how or whether a substance may cause harm. The toxicity or *dose-response assessment* asks how effects might occur following exposure. It identifies the nature of a substance's toxicity via different exposure routes. These may vary, depending on whether the exposure occurs because of inhalation or by ingestion, and also by how much of a substance is absorbed in the body as well as the effects on cells and on whole organisms from these doses. The dose-response assessment asks what is known about the mechanisms of action of substances; how they behave when absorbed (are they metabolized or excreted?); and what are the kinetics, or time-associated parameters, of those reactions in the body. Sometimes substances themselves are not toxic, but they are metabolized to toxic compounds in specific organs in the body. Thus, a detailed understanding of those behaviors is required for dose-response assessments.

The dose-response assessment relies on data from toxicology studies in laboratory animals and studies in test tubes or from those in exposed populations. The dose-response assessment identifies the health effects observed at different doses of substances and determines the lowest levels where effects, or no effects, have been observed. These effects levels become the basis for comparing the exposure levels to effects levels.

Things work differently for substances that are shown to cause cancer in studies or that are toxic to DNA (genotoxic), the basic building block of life. The cancer model currently used assumes that if tumors are observed in test animals at high doses, then the risk to people from low doses in the environment can be extrapolated by drawing a straight line from the high dose to the low dose, adjusting for differences in physiology between laboratory animals and humans. As knowledge of cancer mechanisms improves, other approaches to dose-response assessments are identified, but for most genotoxic substances, this linear low-dose extrapolation method is currently used. The U.S.EPA and others recognize that some substances have different mechanisms of toxicity that may result in different dose-response relationships, but few of these assessments have been carried out to date.

Linear low-dose extrapolation assumes that there is always a probability that cancer might occur: The risk is never zero; it just gets smaller and smaller as exposure levels decrease. For example, a toxicology study might identify the percentage of tumors observed in twennty out of fifty animals (40%) at a high dose, which when extrapolated to environmental levels people could be exposed to might correspond to a cancer risk of 0.00001% for the people exposed! With this methodology, the assumption is that the probability that cancer can occur from exposure is never zero: It continues to lower levels of risk, so low that people generally agree that they are acceptable, or de minimus. This is in contrast to non-cancer risk assessment, where if the exposure is below a no-effect level, it is assumed safe.

The fourth step of risk assessment is *risk characterization*. This step brings together the exposure and dose-response assessments, comparing the exposure levels to the levels associated with health effects, to evaluate whether there is potential for significant risk. Different models and frameworks use different approaches for this step, but generally one can consider the ratio of exposure levels to effects levels. There is often a lot of uncertainty associated with this comparison; the risk characterization step identifies and evaluates these uncertainties.

In the absence of adequate data for comparing exposure levels to toxicity levels, the risk characterization may be qualitative—that is, no numbers are assigned, and the risk level is descriptive rather than quantitative. Also, most of the time, studies are not directly applicable; for example, a dose-response assessment may be based on animal toxicology, and this must be related to human exposure. Moreover, people differ in their susceptibility to certain substances—by age, genetic variation, or because of other factors such as immune status—and these are accounted for when characterizing risk.

Another example of uncertainty is that, often, exposure cannot be measured exactly, so estimates are made that rely on assumptions and extrapolation, i.e., assuming a person drinks two liters (about half a gallon) of water every day. Not everyone drinks exactly two liters of water per day: Some drink one liter, some three liters, and others drink little, if any, water. Children may drink a lot more. This and other sources of population variability add uncertainty to risk estimates because assuming that some exposure occurs when a person ingests two liters of water per day simplifies reality. There is a distribution of risk levels that gets simplified to a "most likely" number.

Risk characterization also evaluates the risks in context of regulations that (a) define how much exposure is allowed under different circumstances and (b) make comparisons with other types of risks that help to inform how the risks are managed. Despite these and other sources of uncertainty, risk assessment is a valuable technique for estimating the potential health and environmental impacts of nanoscale materials used in nanoscience and technology. Even when all of the necessary information is not available, risk assessment can still be helpful to make estimates of potential risk to set research agendas or make safety decisions about working with or using specific materials.

2.7 Issues in Applying the Four Steps of Risk Assessment to Nanomaterials

Because of the unique properties of nanomaterials, particular issues arise in applying the steps of risk assessment specifically to them (Shatkin et al. 2010).

2.7.1 Hazard Assessment

Hazard-identification questions for nanomaterials may not differ from other substances; however, because of the unique properties of nanoparticles, they may need to be measured differently, and the methods for these measurements are not well established. Typically, substances are reported in terms of their concentration, say in "parts per million" (mass of a substance per unit of volume in air, food, water, or blood, for example). But for nanomaterials, the surface characteristics appear to be important, and at the moment, the key properties that affect health are poorly understood. For nanomaterials, the mass concentration may not be the best measure for characterizing risk. Other properties that have not traditionally been measured, such as the number and type of particles or surface properties (surface area, charge, and level of contamination) may be key parameters, but these are not definitively known. Several expert groups have developed lists of the important physical and chemical parameters to measure for assessing and reporting toxicity (see Table 2.1). As Powers et al. (2006) pointed out, even these measures may oversimplify reality.

Some nanomaterials are in groups that cover a range of particle sizes, and this can be difficult to measure accurately and to characterize. The size distribution is important to understanding the behavior, and it adds a layer of complexity, in that many nanomaterials are mixtures of particle sizes and impurities. Another concern is that some particles exhibit dynamic behavior; that is, their properties (the level of aggregation, for

TABLE 2.1

Comparison of Physical Chemical Properties for Nanomaterial Testing

Characteristic	ISO (2012)	OECD (2010)	MINChar (2008)
Agglomeration state/aggregation	X	X	X
Particle size/distribution	X	X	X
Composition	X	X	X
Shape	X		X
Solubility/dispersibility	X		X
Specific surface area	X	X	X
Surface chemistry	X	X	X
Surface charge	X	X	X
Porosity		X	
Crystal structure		X	X
Dustiness		X	
Electron microscopy		X	
Photocatalytic activity		X	
Octanol water partition coefficient		X	
Redox potential		X	
Radical formation potential		X	
Purity			X
Stability			X

example) may change with ambient conditions or other physicochemical changes over time.

2.7.2 Exposure Assessment

For nanotechnologies, some additional issues are raised for characterizing exposure. There may be a need for new metrics to characterize how people are exposed to nanoscale materials. The current techniques available for analysis may not be sensitive or specific enough to understand exposure. Historically, risk assessment has considered the mass or a concentration of a substance. However, with nanoparticles, far more salient features may be the number of particles, the total surface area, and the reactivity of the surface area. New analytical techniques with low detection limits are required; because exposures may be very low, background exposures to nanoparticles may become a major interference. For nanoparticles, there is a need to develop measures to assess and characterize background exposure to nanoscale materials and particles.

With nanoscale materials, it is also not clear how to measure exposure as it relates to toxicity (as a concentration, by surface area, or surface chemistry), and there are few analytical techniques currently available that quantitatively measure substances at the nanoscale. Information is lacking about the transport, fate, and transformation of nanoscale materials in the environment. There is limited understanding of what happens when these particles are released to the environment (Shatkin and North 2010) or even whether they are released. For example, while there are several studies now evaluating what happens to nanoparticles in water used when released to a waste treatment facility, the studies are in model systems and suggest different pathways (see Chapter 5). When released to the environment, are nanoparticles still at the nanoscale? This is an issue in exposure assessment for which there is a need to develop and test models to improve the current understanding. Limbach et al. (2008) tested removal of cerium oxide in a model wastewater treatment plant and found that most nanoparticles were removed by adsorption to the sludge, but that up to 6% of the cerium oxide exited the model plant in the effluent. The authors noted the effects of surfactants in stabilizing the nanoparticles against adsorption onto the sludge.

There is also a need to understand nanoparticle exposure in the body. If you touch nanoparticles, can they penetrate the skin? There is some evidence that because they are so small, some particles can enter the skin. But, can they get across the outer barrier layer, the stratum corneum, and enter the body? Similarly, if particles are inhaled, are they absorbed by the lung and can they cross from the lungs into the blood? If they enter the nasal cavity by nose breathing, can they travel into the nervous system? What happens to engineered nanomaterials that are eaten? Do we excrete them, absorb them into the body, or do they stay and accumulate in the liver? These issues are explored in greater depth in Chapter 4, but they are important areas of exposure assessment that need to be addressed for evaluating exposure to nanoscale materials.

Today, it is likely that any exposures to engineered nanoparticles are very low. For most nanoparticles, production levels are below reporting thresholds, indicating a low potential for widespread consumer and environmental exposure. Consequently, now is the time to develop exposure-characterization methods, before large-scale production occurs. And it is wise to consider the many pathways of exposure to nanomaterials in the products we might use over their entire life cycle, from their creation, manufacturing into products, use by consumers and industry, and disposal. By considering these exposures now, we can minimize risks that have been anticipated.

2.7.3 Dose-Response Evaluation

As discussed in Chapter 4, one widely observed effect from exposure to nanoparticles is inflammation, an immune system response. Inflammation is associated with the development of many diseases such as asthma, heart disease, cancer, and autoimmune diseases. Inflammation has been observed following exposure to nanomaterials in whole-animal studies (*in vivo*) and in cellular assays (*in vitro*). It is presently unclear whether or in what ways the chemical composition, size, shape, or surface characteristics affect the toxicity of nanoscale materials.

Classical models of toxicity look at a range of noncancer end points, including whole-system toxicity, reproductive end points, neurological effects, and effects on the immune system. Carcinogenic effects (the ability of the substance to cause cancer) are also commonly evaluated for substances. Again, there is uncertainty in defining the dose. Dose-response studies have demonstrated that some types of nanoparticles behave differently than larger particles, and this may create difficulties in measuring responses. Also, studies have demonstrated that small changes, such as the crystalline structure and the surface charge of a nanoparticle, can greatly affect its behavior in the body. The outcome of the dose-response assessment is to identify the lowest levels at which an effect occurs (the lowest observed adverse effects level, LOAEL) or the level at which no effects occur (no observed adverse effects level, NOAEL). One complicating factor is the reactivity of the surface of nanoparticles. Because they have large surface area relative to their size, the surfaces of nanoparticles tend to bind to other molecules. So, they can form a surface of proteins or fat molecules, depending on where they are introduced. The subsequent behavior, including toxicity, is influenced by this binding activity, and it is poorly understood.

A number of researchers are trying to develop predictive approaches to toxicity studies that do not involve testing in whole animals. These test tube or *in vitro* (literally, in glass) assays to date have not been shown to be relatable to *in vivo* studies for nanoparticles. This may be due to the difficulties of getting nanoparticles to separate from one another. Most nanoparticles are very sticky and tend to agglomerate or aggregate. When they do, it is difficult to measure the toxic effects or to be certain about the exposures that have occurred. Some researchers have addressed this by using techniques to separate the particles, so it then becomes a question of whether the findings can be related to real-world exposures. One recent study showed that some nanoparticles interfere with the reactive agents in *in vitro* assay, producing a false positive (the assay would say there was an effect when there was not). Many hurdles remain for assessing the toxicity of nanomaterials. These are explored in more depth in Chapter 4.

2.7.4 Risk Characterization

Risk characterization asks, "What does the assessment infer about health, safety, and the environment?" Considering risks from nanomaterials and nanotechnologies in context may mean comparison to existing standards for risks posed by other substances for which there has been greater investigation. There are few available risk assessments for nanomaterials, and they are generally based on mass concentration, which is a limitation. These are discussed in Chapters 5 and 6.

An additional challenge is that most people will come closest to exposure with nanomaterials from using them in products, as opposed to those who generate the raw materials in a work environment and may be exposed to the nanoparticles themselves. In the workplace, exposure to substances can be managed if there is determination to do so—it is a fairly controlled environment. But in the broader environment, it may be the products that need to be managed, not the substances, because their uses will vary greatly. In some cases, users may never know there were nanomaterials in the product. Determining the potential risks associated with the use of nanomaterials in products is an added challenge for risk management and risk assessment. In Chapter 3 and beyond, we consider how risk assessment can help accomplish this task while achieving the many benefits of nanotechnologies without incurring so many of the risks.

References

Anon. 2007. A quick history of bicycles. Pedaling History Bicycle Museum. http://www.pedalinghistory.com/PHhistory.html

Asian Economic News. 2001. Thai consumers get "Frankenstein food," Greenpeace says. April 16, Kyodo News International.

ASTM International. 2004. Standard guide for risk-based corrective action. E2081-00(2004)e1. http://www.astm.org/DATABASE.CART/HISTORICAL/E2081-00R04E1.htm

CDC. 1999. Revised final FY 1999 performance plan and FY 2000 performance plan. http://www.cdc.gov/program/performance/fy2000plan/pp99.pdf

Creighton, M. 2002. *Prey.* New York: Harper Collins.

Curall, S. C., E. B. King, N. Lane, J. Madera, and S. Turner. 2006. What drives public acceptance of nanotechnology? *Nature Nanotechnol.* 1(3): 153–55.

EPA. 1998. Guidelines for ecological risk assessment. EPA/630/R095/002F. Risk Assessment Forum, U.S. Environmental Protection Agency, Washington, DC.

———. 2004. An examination of risk assessment principles and practices. Staff paper. Office of the Science Advisor, U.S. Environmental Protection Agency, Washington, DC. http://www.epa.gov/osa/pdfs/ratf-final.pdf

European Commission. 2006. REACH. http://ec.europa.eu/environment/chemicals/reach/reach_intro.htm

FAO. 2005. Understanding the Codex Alimentarius. http://www.fao.org/docrep/008/y7867e/y7867e00.htm

FDA. 2001a. HACCP: A state-of-the-art approach to food safety FDA Backgrounder. BG 01-4. http://www.scribd.com/doc/78555546/FDA-Backgrounder-HACCP-a-State-Of-The-Art-Approach-to-Food-Safety

———. 2001b. HACCP overview. U.S. Food and Drug Administration. http://www.fda.gov/Food/FoodSafety/HazardAnalysisCriticalControlPointsHACCP/default.htm

Finkel, M. 2007. Malaria: Bedlam in the blood. *National Geographic* 212(1): 32–67.

ISO. 2012. Nanotechnologies: Guidance on physico-chemical characterization of engineered nanoscale materials for toxicologic assessment. ISO/DTR 13014 (draft).

Kahan, D., P. Slovic, D. Braman, J. Gastil, and J. L. Cohen. 2007. Affect, values, and nanotechnology risk perceptions: An experimental investigation. Working Paper Series. Social Science Research Network. http://ssrn.com/abstract = 968652

Kates, R. W., C. Hohenemser, and J. X. Kasperson. 1985. *Perilous progress: Managing the hazards of technology.* Boulder, CO: Westview Press.

Limbach, L. K., R. Bereiter, E. Muller, R. Krebs, R. Galli, and W. Stark. 2008. Removal of oxide nanoparticles in a model wastewater treatment plant: Influence of agglomeration and surfactants on clearing efficiency. *Environ. Sci. Technol.* 42(15): 5828–33.

MINChar. 2008. Characterization matters: Recommended minimum physical and chemical parameters for characterizing nanomaterials on toxicology studies. Minimum Information for Nanomaterial Characterization (MINChar) Initiative. http://characterizationmatters.org/parameters/

Morgan, M. G., P. Slovic, I. Nair, D. Geisler, D. MacGregor, B. Fischhoff, D. Lincoln, and K. Florig. 1985. Powerline frequency electric and magnetic fields: A pilot study of risk perception. *Risk Analysis* 5(2): 139–49.

Nanotechweb. 2006. Nanotechnology "more risky than bicycles." http://nanotechweb.org/articles/news/5/12/3?alert = 1

———. 2006b. IOP Publishing. http://nanotechweb.org/cws/home

NIOSH. 2007. Draft NIOSH current intelligence bulletin: Evaluation of health hazard and recommendations for occupational exposure to titanium dioxide. NIOSH Docket No. 100. http://www.cdc.gov/niosh/review/public/TIo2/default.html

NRC. 1983. *Risk assessment in the federal government: Managing the process.* Washington, DC: National Academy Press.

———. 1996. *Understanding risk: Informing decisions in a democratic society.* Washington, DC: National Academy Press.

———. 2009. *Science and decisions: Advancing risk assessment.* Washington, DC: National Academy Press.

OECD. 2007. Environment directorate. Chemical hazard/risk assessment. Organization for Economic Cooperation and Development, Paris. http://www.oecd.org/department/0,2688,en_2649_33713_1_1_1_1_1,00.html

———. 2010. OECD Environment, Health and Safety Publications Series on the Safety of Manufactured Nanomaterials No. 27. List of manufactured nanomaterials and list of endpoints for phase one of the sponsorship programme for the testing of manufactured nanomaterials (revision). ENV/JM/MONO (2010) 463. Organization for Economic Cooperation and Development, Paris.

Petty, R. D. 1991. Regulation vs. the market: The case of bicycle safety. Part I, 2.77, and Part II, 2.82. http://www.piercelaw.edu/risk/vol2/winter/petty1.htm

Powers, K. W., S. C. Brown, V. B. Krishna, S. C. Wasdo, B. M. Moudgil, and S. M. Roberts. 2006. Research strategies for safety evaluation of nanomaterials. Part VI: Characterization of nanoscale particles for toxicological evaluation. *Toxicol. Sci.* 90(2): 296–303.

Renn, O. 2005. Risk governance: Towards an integrative framework. White paper No. 1 International Risk Governance Council. http://www.irgc.org/IMG/pdf/IRGC_WP_No_1_Risk_Governance_reprinted_version_.pdf

Renn, O., and Roco, M. 2006. Nanotechnology Risk Governance. White paper. International Risk Governance Council. http://www.irgc.org/IMG/pdf/IRGC_white_paper_2_PDF_final_version-2.pdf

Robichaud, C. O., D. Tanzil, U. Weilenmann, and M. R. Wiesner. 2005. Relative risk analysis of several manufactured nanomaterials: An insurance industry context. *Environ. Sci. Technol.* 39(22): 8985–94.

Shatkin, J. A. 2008. Informing environmental decision making by combining life cycle assessment and risk analysis. *J. Ind. Ecol.* 12(3): 278–81.

Shatkin, J. A. 2009. Advancing risk communication: Uncertainty and the NRC report. *EM* 2009(7): 22–25. http://pubs.awma.org/gsearch/em/2009/7/shatkin.pdf

Shatkin, J. A., L. C. Abbott, A. E. Bradley, R. A. Canady, T. Guidotti, K. M. Kulinowski, R. E. Lofstedt, et al. 2010. Nano risk analysis: Advancing the science for nanomaterials risk management. *Risk Analysis* 30(11): 1680.

Shatkin, J. A., and W. North. 2010. Perspectives on risks of nanomaterials and nanotechnologies: Advancing the science. Introduction to the special series. *Risk Analysis* 30(11): 1627.

Simons, J., R. Zimmer, C. Vierboom, I. Härlen, R. Hertel, and GbF. Böl. 2009. The slings and arrows of communication on nanotechnology. *Nanopart Res.* 11(7): 1555–71.

Slovic, P., Fischhoff, B., & Lichtenstein, S. (1985). Characterizing perceived risk. In R. W. Kates, C. Hohenemser and J. Kasperson (Eds.), Perilous progress: Managing the hazards of technology. (pp. 91–125). Boulder, CO: Westview.

Starr, C. 1969. Social benefit versus technological risk. *Science* 165(3899): 1232–38.

U.S. Presidential Commission on Risk Assessment and Risk Management in the Federal Government. 1997. Framework for environmental health risk management. Final report, Vol. 1. Washington, DC. http://www.riskworld.com/nreports/1997/risk-rpt/pdf/EPAJAN.PDF

Weiss, R. 2006. Nanotech product recalled in Germany. *Washington Post*, April 6, A02. http://www.washingtonpost.com/wp-dyn/content/article/2006/04/05/AR2006040502149.html

WHO. 2004. IPCS risk assessment terminology. Part1: IPCS/OECD key generic terms used in chemical hazard/risk assessment; Part 2: IPCS glossary of key exposure assessment terminology. World Health Organization, Geneva, Switzerland. http://www.who.int/ipcs/methods/harmonization/areas/ipcsterminologyparts1and2.pdf

Further Reading

Davis, J. M., A. Wang, J. A. Shatkin, J. A. Graham, M. Gwinn, and B. Ranalli. 2009. Nanomaterial case studies: Nanoscale titanium dioxide in water treatment and in topical sunscreen (external review draft). National Center for Environmental Assessment, Office of Research and Development, U.S. EPA, Research Triangle Park, NC.

Oberdörster, G., A. Maynard, K. Donaldson, V. Castranova, J. Fitzpatrick, K. Ausman, J. Carter, et al. 2005. Principles for characterizing the potential human health effects from exposure to nanomaterials: Elements of a screening strategy. *Particle Fibre Toxicol.* 2: 8.

3

Sustainable Nanotechnology Development Using Risk Assessment and Applying Life Cycle Thinking

Jo Anne Shatkin

CONTENTS

For a successful technology, reality must take precedence over public relations, for nature cannot be fooled.

Richard P. Feynman

To reach nanotechnology's full potential over the next decade, it is vital to combine economic support with meaningful incentives and frameworks to ensure responsible development that, besides technological and business goals, also addresses societal goals.

Mikhail Roco

Technology is as old as history itself, and for centuries has been at the core of our economy. In the machine age, the steam engine radicalized transportation, and ever since, we've been innovating in the ways we use natural resources to meet our societal needs and desires. Science and engineering studies have identified and developed a wide array of materials with specific properties useful for economic development.

Yet, we do not need to look far to find examples of how past approaches to managing technological risk were inadequate. Perfluorinated organic compounds, lead, PCBs, and asbestos—each are examples of materials that have an environmental legacy that proved costly. Engineers and materials scientists found that they possessed unique and brilliant properties, and identified more and more applications for them. These materials were developed or used widely, while their toxic properties were not identified until later, in part because they offered benefits as well.

Nonstick polymer coatings are frequently used to coat cookware, packaging materials, fabrics, and medical equipment; they reduce friction, increase water resistance, retard flammability, and limit staining potential. As a polymer, these coatings have little to no toxicity when the compounds used to make them are tightly bound in the matrix. But when heated to high temperature, the building blocks of the polymer can release into the air. Some of these building blocks—perfluorinated organic compounds—have been found to be very persistent in the environment. Persistent compounds are of environmental concern because they bioaccumulate; that is, low levels in the environment tend to increase in concentration as they move up the food chain, for example from water to small fish to bigger fish to fish-eating birds and wildlife. Perfluorinated compounds have been measured in people's blood in countries all over the globe as well as in drinking water, fish, birds, and marine mammals, including polar bears off the coast of Greenland (Bossi et al. 2005; WWF 2004). Some of these compounds (e.g., perfluorooctane sulfonate [PFOS]) are no longer manufactured, and U.S.EPA is conducting a risk assessment of several of the perfluorinated compounds.

Lead was used in antiquity in glazes, paints, beads, currency, and cosmetics, and it is mentioned in the Bible. Lead poisoning occurred in Roman times, including the effects from lead in pipes. Lead poisoning was widespread during the Industrial Revolution. In 1925, lead was the first chemical contaminant to be regulated in drinking water in the United States, long before the establishment of the U.S.EPA. Yet lead possesses unique properties that found wide application in fuel stabilizers, pesticides, pigments, soldering pipes, automotive batteries, bullets, insulation, and glass (Winder 2004; ATSDR 2005).

The U.S.EPA still spends millions of dollars of its shrinking annual budget on lead abatement. Lead is an elemental metal, and it cannot be further broken down. At one time, lead was used in gasoline, paints, and plumbing systems. Even though it is now banned from these uses, people are still exposed to lead, directly or indirectly, as a result of its prior use. Children still suffer from lead poisoning, mainly from exposure to lead paint. After years of exhaust emissions from automobiles that spread lead—first into the air, then to soils—those now living near major roadways may have lead in the food they grow. When the wind blows, lead in the soil can be entrained in the air and enter people's nearby homes or workplaces. This lead and other pollutants then become part of the dust that people inhale and that young children can ingest while playing on the floor.

There is an *opportunity cost* associated with lead abatement, the missed opportunity of investing in something else, such as research on new technologies that provide clean drinking water. But the existing environmental and health burdens from past activities require at least the current level of effort to reduce lead levels in the environment and to find ways to prevent children and others from exposure to lead in their homes, drinking water, toys, and play areas.

Numerous lead-abatement measures in the United States have been enacted since the 1970s, and lead levels have decreased significantly. Lead levels in air have decreased to 5% of 1980 levels, following the banning of lead from gasoline. The number of children in the United States with elevated blood lead levels is 97% fewer than in 1978. But because lead is an element and does not break down, regulatory action is still required to prevent exposure to painted building materials, plumbing, and soils that contain lead (U.S.EPA 2007).

3.1 Fuel Additives: The Costs of Lack of Foresight

While lead levels in air are decreasing, lead in urban soils remains a concern. When lead was banned from fuels, its main replacement was an additive called MTBE (methyl *tert* butyl ether), another relevant and interesting example of the unsuspected environmental impacts created by our lack of foresight. MTBE is an oxygenate, a fuel additive that promotes cleaner burning of gasoline, increases octane, and reduces automobile emissions; it was developed to reduce air pollution from carbon monoxide and ozone (Davis and Farland 2001; CalEPA 1999; U.S.EPA 2006). The 1990 Clean Air Act amendments mandated that oxygenates (not MTBE specifically) be added to gasoline to improve air quality in areas of high air pollution in the United States. MTBE was not considered very toxic when its use was approved for reformulated gasoline. Problems soon arose, however. The first reports of adverse health effects were complaints of headaches and nausea following exposures at filling stations.

MTBE is now banned from gasoline. The main reason was its wide-spread release to the environment. While intended to improve air quality, MTBE emerged as highly mobile and very soluble in water, often at the leading edge of other petroleum constituents that leak into groundwater (water that flows underground). When gasoline leaks—from an under-ground storage tank or from an accidental spill during fuel transfer—it runs underground until it reaches groundwater; and once in this medium, MTBE can move very quickly and contaminate drinking water supplies.

When MTBE was banned in California in 1998, the state had already set a public health goal for drinking water based on its cancer-causing potential, although it remains controversial whether MTBE is a human carcinogen (cancer-causing substance) (NRC 1996; Stern and Tardiff 1997; Mennear and McConnell 1997; Mehlmann 2002). Some researchers claim that studies of carcinogenicity in rats and mice are not relevant to humans because of (1) differences in biology, and (2) the very high dose levels asso-ciated with the effects in the studies, which humans would never toler-ate because of taste and odor issues. Others claim rodent carcinogenicity models are inadequate predictors for low-dose human exposure (Ames and Gold 2000). As of July 2012, the U.S.EPA has not completed a deter-mination of the carcinogenic potential of MTBE, and has not regulated MTBE in drinking water, although the U.S.EPA has prioritized MTBE for consideration, listing it on the Contaminant Candidate List, the U.S.EPA's list of priorities for regulatory determination for drinking water stan-dards (U.S.EPA 2009).

As discussed in more detail by J. Michael Davis (formerly of the U.S.EPA) in Chapter 7, in their alternative fuel strategy over a decade ago, U.S.EPA called for "research to assess the impact of reformulated gasoline on the potential for groundwater contamination and result in pollutant exposure" and to "characterize the impacts of oxygenates on the fate and transport of fuel components" (U.S.EPA 1992). However, the early warn-ing signs of toxicity and environmental mobility were not heeded soon enough, and now MTBE has a legacy that many property owners, oil com-panies, and drinking water suppliers must address. MTBE, the safer alter-native to lead, became its own significant environmental issue.

Davis and Farland (2001) point out that the MTBE experience should teach us to look comprehensively at the behavior of any additives that we might introduce into fuels or broadly into the environment. There were early signs that MTBE was not a good candidate for widespread introduction in the environment. The chronology of MTBE's rise and fall is also instructive: Oxygenates were mandated for use in gasoline in 1990 and in reformulated gasoline in 1995. In 1997, MTBE was the second most produced chemical in the United States, but was banned first in California in 1998, and it is now banned in the United States. Among other things, the MTBE example demonstrates that society is collectively learning to act and react relatively more quickly to substances that pose risks. However, this experience teaches that there are benefits to being more proactive in managing and more broadly considering the risks as technologies are developed and before they are commercialized.

3.1.1 Nanoscale Cerium Oxide, Today's Diesel Fuel Additive

These examples demonstrate that there is a cost to introducing materials into the environment without considering their long-term impacts. These *opportunity* costs are also risk-risk trade-offs, and it makes sense to consider them early in technology development. Research for lower impact materials could be funded if less were spent on managing legacy pollutants.

Nanoscale cerium oxide is now being used commercially for a variety of uses, including as a fuel additive that improves fuel burning and reduces the release of particles from the exhaust of diesel engines. Since 1999, cerium oxide has been used as a diesel fuel additive in the United States (Cassee et al. 2011), and it underwent trials in the United Kingdom in 2004–2006 and is now used there as well. Adding cerium oxide nanoparticles to diesel fuel has been demonstrated to improve fuel efficiency while reducing soot formation (lowering air pollution) and volatile emissions (Energetics 2012; Cassee et al. 2011). In 2001, the Health Effects Institute (HEI) conducted a risk assessment of cerium oxide in fuels, concluding that:

> Based on the limited data available, toxicity of cerium oxide appears to be small, and cerium oxide might not be of concern when inhaled or ingested at the low levels that would be encountered in the environment.... The absence of more complete information precludes fully assessing the possible health effects of using cerium as a fuel additive.... Considerations are the additive's ability to reduce harmful emissions, its persistence in the environment, and the feasibility and cost effectiveness of this technology in comparison with other technologies that can achieve these reductions. (HEI 2001)

Now applying the lens of the emerging understanding of nanoparticle exposure and effect, in a more recent review, Cassee et al. (2011) concluded, "The potential environmental, health and ecological effects of cerium nanoparticles remain uncertain and require additional research." The authors expressed concerns about the dispersive nature of cerium oxide exposure as a fuel additive and about the limited data for risk assessment, which were derived from controlled exposure studies that do not reflect real world conditions, and about the effects data, which were mainly from *in vitro* toxicology studies.

There are clear health and environmental benefits from adding cerium oxide to diesel fuel, and the available data suggest that the risks appear to be low. However, the release of cerium oxide in vehicle exhaust is increasing the environmental concentrations of cerium, and the pathways to other types of environmental exposure require consideration. The broader issues raised by HEI and Cassee et al. (2011) about cerium oxide can be addressed in a broader, risk-informed framework.

A risk assessment of nano cerium oxide (nano-CeO_2) was conducted by a U.K. manufacturer. The assessment relied on *in vitro* cellular studies of the effects of nano-CeO_2, including comparative studies to conventional CeO_2, and found no suggestion of toxicity at environmentally relevant levels. The results suggest greater health benefits than risks, due to the decrease in particulate matter (Park and Martin 2009).

The proactive nature of this risk assessment is a step forward, as was the early assessment of MTBE risks (described in Chapter 7). Nevertheless, the existing studies for nano-CeO_2, the current generation of fuel additives, are inadequate for risk assessment. Several studies suggest that *in vivo* toxicity of nanoparticles is not currently predictable from *in vitro* assays (see Chapters 4 and 5). Thus, despite indications of nano-CeO_2 being low toxicity, given the trajectory for widespread release of nano-CeO_2 from diesel bus and truck emissions, the unanswered questions about *in vivo* toxicity and environmental behavior of nano cerium oxide in diesel fuel exhaust (Cassee et al. 2011; Cho et al. 2010) need to be addressed now to ensure that we understand the risks before we create an environmental burden with the next generation of fuel additives.

3.2 Risk Assessment for Nanotechnology Is Urgently Needed

We noted in Chapter 1 several factors that make it imperative to begin the assessment of risk for nanotechnology now, even without available data— including its rapidly accelerating pace of development, lack of adequate information about nanomaterials, their potential for wide dispersion in the environment, and the lack of standards or guidelines regulating them. These are explored in more depth in the following subsections.

3.2.1 The Pace of Nanotechnology Development and the Paucity of Information

The accelerating pace of development described in Chapter 1 poses a challenge for health and safety research and the characterization of risks. With thousands of patents in nanotechnology, and well over 1,000 specifically for nanomaterials, more research on health risks is needed. Each formulation and manufacturing process creates slight differences in size, surface area, aggregation state, and contamination level. Understanding the impacts of these differences is an important factor in understanding risks.

At first glance, a lot of information seems to be available: hundreds of studies evaluating toxicology, behavior in cell systems, and behavior in the environment for select nanomaterials. It is clear from examining these papers, however, that there is a lot of uncertainty about the key parameters affecting toxicity. Increasingly, the level of sophistication of these studies improves, yet there are still many studies where the physical and chemical properties are not adequately characterized. Most studies investigate a relatively small set of materials, and some suggest that the test system may also affect the findings, particularly for *in vitro* studies, and there is still difficulty in getting reproducible results in laboratory tests. The same materials can produce different results in different tests. As discussed in Chapters 4 and 5, the results of toxicity testing can be equivocal—positive in one study and negative in another—and it depends on how the materials were prepared and handled as well as the test conditions. The example of cerium oxide as a fuel additive is instructive. The existing data are inadequate for risk assessment.

While there are many reports, they do not point to a clear understanding about the toxicity of nanomaterials, such that as of this writing, it is not possible to conclude that nanomaterials are safe or toxic. The OECD (Organization for Economic Cooperation and Development) member-sponsored studies now underway may provide some answers soon. Answers, however, are needed now to guide decision making about occupational and environmental protection.

3.2.2 Potential for Wide Dispersion in the Environment Amid Uncertainty

Products that consumers use are, by definition, widely dispersed. They go wherever people that use them go. If a product is sold in Wal-Mart, for example, it will travel to the twenty-seven countries in which Wal-Mart operates (walmart.com). With nanotechnology applications in lotions, creams, cosmetics, sporting goods, food packaging, paints, fabrics, electronics, and automobiles, nanomaterials are and will be widely dispersed geographically. Some applications have greater potential than others for actual release into the environment (as we have seen, this is a factor of exposure), but all have the potential for global distribution, even with variances in local markets.

3.2.3 Few Standards or Guidelines

The regulatory situation is dynamic and is discussed in greater detail in Chapter 10, Ongoing International Efforts to Address Risk Issues for Nanotechnology. It has certainly changed since the first edition of this book was published in 2008. At that time, only one city, Berkeley, California, had a nanomaterial-specific regulation, requiring all users to register and provide the city with information about the materials they were using.

One reason Berkeley developed a nanotechnology regulation is that there were none existing at the federal level. The U.S. federal regulatory process is slow. Even if efforts began today to regulate nanotechnology, it could be decades before they were enacted (Brown et al. 1997). U.S. governmental agencies have taken steps to regulate the new substances that have crossed their thresholds for regulatory approval. The efforts initially were on a case-by-case approach, where regulators worked with the submitter to obtain and review key information. In the case of the Toxic Substances Control Act (TSCA), U.S.EPA has developed policies for new chemical submissions that require specific information about nanomaterials, and it has reviewed over 100 submissions through 2011. As described in Chapter 10, U.S.EPA has published several Significant New Use Rules regarding specific nanomaterials.

An issue is that there are different regulatory requirements for different applications of nano-enabled products: Regulatory approvals for drugs are different from cosmetics; chemicals are different from foods; pesticides are different from food contact substances. Electronics and consumer products are among the categories of not requiring "premarket approval," and these are regulated only if problems are identified after commercial introduction.

Some existing regulations at the federal level do apply to nanotechnology. For example, the General Duty Clause under OSHA (Occupational Safety and Health Administration) requires that "each employer shall furnish to each of his employees employment and a place of employment which are free from recognized hazards that are causing or are likely to cause death or serious physical harm to his employees" (OSHA Act 1970).

Some nonregulatory organizations in the United States have proposed guidelines for occupational handling of nanomaterials. As described in Chapter 8, the National Institute for Occupational Safety and Health *"Approaches to Safe Nanotechnology"* (NIOSH 2005, 2009) describes "what is known about nanomaterial hazard and measures that can be taken to minimize workplace exposures." A collaborative effort developed a "Good Nano Guide" wiki (ICON 2010) intended for researchers to share best practices and policies for managing nanomaterials in occupational environments. Unfortunately, it remains in beta form, waiting for new leadership. There are also efforts to develop voluntary standards for occupational environments, such as at ASTM International and ISO. These are discussed in more detail in Chapter 10.

3.3 Environmental Risk Issues

In the previous examples, problems arose when materials that posed risks were released into the environment and caused exposures to people that resulted in toxicity, eventually resulting in action to mitigate those concerns. Perfluorinated compounds, lead, and MTBE will be in the environment for some time to come and will require continued attention. As previously discussed, the opportunity costs associated with managing and removing lead from the environment consume resources that could be used elsewhere. The costs of environmental remediation also do not create future economic opportunities. These contaminants are mostly of concern because they affect people. What about effects on the rest of the ecosystem?

Figure 3.1 shows how contaminants can move into a water body. Ecosystems are usually very complex, and the dynamics are generally poorly understood. Risk models generally assume that each environmental contaminant behaves individually, regardless of what else is present. There are so many combinations, it could consume all of the public health resources of the world to understand the complex interrelationships, so assessments generally assume that there is no relationship. The risk from one exposure is simply added to the next in terms of impact. But in reality, exposures are due to a multitude of combinations of substances in the environment.

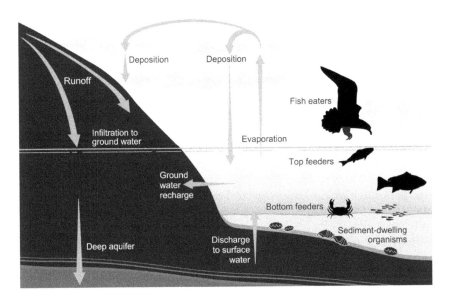

FIGURE 3.1
Movement of contaminants into a water body.

Presumably, the behavior of nanomaterials is going to vary, depending on the material and the type of ecosystem. Is there a need to worry about affecting the ecosystem by releasing poorly characterized materials into it? Some newly emerging concerns include the release of hormones into the environment. Every time a toilet is flushed, hormones, whether naturally occurring or synthetic, are sent out into the environment. Many people now take prescription drugs, including female and male hormones, such as steroids, estrogen, and progesterone. The U.S. Geologic Survey conducted a national study of the occurrence of hormones and other wastewater contaminants and measured low levels downstream of wastewater treatment plant effluents (Kolpin et al. 2002). In urban rivers and streams, fish and other organisms are experiencing the effects of these hormone exposures, with a shocking prevalence of hermaphroditic fish containing both male and female characteristics (Jobling et al. 2006). Because the levels are low (water dilutes the concentration), many people do not feel it is important to address this concern. This cycle of life returns to humans, as people consume plants, animals, and fish. Humans are a part of the ecosystem, so even from a resource perspective, it is critical to protect the environment. It is important to our future that we do not pollute our own backyard.

3.3.1 Environmentally Friendly Nanotechnology

Inventors, chemical engineers, and materials scientists identify and create materials as means to achieve specific properties. Some call them nanoscientists and nanotechnologists. The discovery and exploitation of specific behavior is behind the development of novel materials with increased reactivity, conductivity, optical properties, flexibility, strength, and thermal properties—a host of desirable properties.

However, materials science is far removed from environmental science. No part of a chemical engineer's curriculum trains students on how materials behave in the environment; there is no discussion of the types of materials that become water pollutants or are persistent in the air, potentially contributing to smog, global warming, or asthma attacks. Simply, inventors are not trained to think about the downstream or long-term effects of the materials and new technologies they create. Rarely are new materials sought because they are less toxic, less persistent, or less biologically available.

There is a growing movement toward so-called green nano or sustainable nanotechnology that seeks to reduce the toxicity and environmental burden associated with substances and technologies, including a series of meetings on the topic. Generally, certain material properties are tested as indicators of potential for persistence, potential to accumulate in the food chain (bioaccumulation), or toxicity. Several organizations (e.g.,

U.S.EPA and the European Union) fund research into models that evaluate the behavior of materials across their life cycle, from their creation or extraction from the earth to their post-use fate in the environment, whether by recycling or disposal. U.S.EPA also has programs, such as Design for the Environment, that seek out chemicals of lower toxicity to replace toxic chemicals currently used in manufacturing. Life cycle analysis is also popular in Germany, Denmark, and many other nations as well as private corporations. There are many initiatives addressing green technology development. These initiatives generally focus on limiting the hazard side of the risk equation. They generally do not consider how limitations on exposure affect the "greenness" of innovations. The focus of this book is on the ways we can be better informed about technology impacts through risk analysis, because it considers both hazard *and* exposure.

3.4 Risk Assessment for Nanotechnology

It is clear that risk assessment is an important tool for understanding and managing the potential risks from nanotechnologies and nanomaterials. Despite existing data gaps and the many questions that still need to be answered, risk assessments can be used as a screening tool that can be applied now to help identify and prioritize the information needed, and to help understand what is not known—but what needs to be known—before making decisions about new technologies. Even with current uncertainties, the risk analysis framework can help to make better decisions today.

Assessments can address both risks and benefits, weighing them against each other. For example, using cerium oxide in fuel improves air quality, a benefit of reducing particulate emissions from diesel exhaust, but this needs to be weighed against the risks of releasing cerium oxide nanoparticles into the air. Risk analysis allows this type of comparative analysis. It is also useful to compare the risks and the benefits of cerium oxide additives with those of other clean-burning alternatives such as ethanol.

3.5 Life Cycle Analysis for Sustainable Nanotechnology

Sustainable development was defined by the Brundtland Commission as "not a fixed state of harmony, but rather a process of change in which the exploitation of resources, the direction of investments, the orientation of technology development, and institutional changes are made consistent with future as well as present needs" (UN 1987). More recently, sustainability was defined as a recognition "that the biosphere, or natural capital, sustains the economy, which in turn supports quality of life" (Koehler and Hecht 2006). Other definitions are less encompassing. Hess (2007), in an analysis of U.S. decisions to switch to cleaner buses, finds that "'sustainability' is not a goal that can be easily defined in the neutral and unbiased way, but a field of contestation that involves ongoing negotiations over fundamental definitions."

As with risk analysis, it is a bit far afield for this discussion to focus on a common definition of sustainable technology development, but the critical contribution is that new and emerging technology and product development considers a broad set of potential societal and environmental impacts, and seeks to understand and minimize these. Sustainable technology development represents a proactive approach to managing risks, by considering them and eliminating them *a priori*, before widely introducing new materials and products into the environment untested. For example, some view sustainable technologies as carbon neutral—that is, they do not contribute to releases to the atmosphere of carbon dioxide, the main greenhouse gas contributing to climate change. The focus of much sustainable (or "clean") technology development is on energy efficiency. Sustainable technology also must address human and environmental impacts, in terms of harm to health and the environment versus improving health and environment. Technology is the development of response to a human need or want. There are choices about technologies that are developed. Why not focus on technologies that "first, do no harm"—the thesis of the medical oath?

This is an idea briefly suggested earlier: that new technologies should convey benefits that our existing technologies do not. However, they also often bring new risks or trade-offs, and it is important to broadly consider the proposed benefits and risks of new technologies in comparison to existing technologies. This is the broad principle—industrial ecology—behind *life cycle analysis* (LCA).

Some refer to industrial ecology as the science of sustainability. Industrial ecology seeks to develop technologies that operate without degrading the environment. LCA is a tool to evaluate whether technologies accomplish that goal. There are many alternative definitions of industrial ecology, and these are not reconciled here. Rather, this discussion focuses on the application of LCA as a study of the flows of materials and energy associated with product development.

Many applications of LCA use *life cycle impact analysis* (LCIA). LCIA applies modeling tools to evaluate impacts of technologies broadly, from large databases for materials from their generation, or removal from the earth, to their end of use. Terms used to describe the life cycle include:

cradle-to-grave, or the beginning to the end of life; cradle-to-gate, the industrial process of developing products; and cradle-to-cradle, William McDonough and Michael Braungard's notion that products should be designed for systems that regenerate, that are ecological by design, and for which the end of life is a new beginning, not a waste product. The materials flowing from one technology become inputs for another, limiting impacts on the environment and on resource depletion (McDonough and Braungard 2002).

LCIA was developed through substantial efforts by the Society for Toxicology and Environmental Chemistry (SETAC) and the International Organization for Standards Technical Committee (ISO 14040) to agree on impacts and criteria for their approaches to life cycle assessment. LCIA is promoted through a variety of voluntary initiatives by the United Nations Environment Programme (UNEP), national governments, industry, and many international organizations and industry associations. Increasingly, LCA is becoming part of industrial and governmental efforts to evaluate products. Several models are currently used for LCIA, but agreement about using them varies depending on who conducts the analysis. LCIA is a key tool of the U.S. Green Building Council's Leadership in Energy and Environmental Design (LEED), a rating system for developing buildings in a sustainable way, often called "green building."

These concepts represent a new way of thinking about technology development that is more holistic than simply identifying a material with new properties and replacing existing technology. LCIA applies an indicator approach: It assesses the inputs and outputs at each stage of the life cycle to estimate the impacts of a product or technology on a set of predetermined criteria. The criteria include: greenhouse gas emissions, air pollutants (nitrogen oxides and ozone precursors), acid rain, smog, impacts on water by the path of eutrophication (overgrowth of algae in water bodies that causes lakes to die), natural resource depletion, and effects on human and environmental species. LCIA models estimate releases of materials that impact each of these categories; the emissions are matched with material properties to estimate an overall impact that can be compared to other products, so that the product with the least impact can be identified.

One idea bred from this work is the notion of *life cycle thinking*, which "integrates existing consumption and production strategies, preventing a piece-meal approach" (UNEP 2001; EU Joint Research Centre 2007). *Life cycle thinking* applies the concepts of considering the broader context for a product in its design and development, without necessarily the detailed data requirements of LCA and LCIA. Some have suggested it as a tool for small and medium enterprises (SME).

My organization, CLF Ventures, uses a tool called Life Cycle Scanning (LC Scan™) to conduct high-level evaluations of technologies and other products as a way to identify hot spots, for example, environmentally dispersive uses that warrant more detailed evaluations using life cycle assessment. LC Scan is a way of explaining the key benefits and risks to stakeholders and informing the critical analyses for addressing those risks.

Life cycle thinking for nanotechnology development is currently promoted in terms of green chemistry. One recent initiative is summarized in *Nanotechnology and Life Cycle Assessment: A Systems Approach to Nanotechnology and the Environment*, which discusses the potential for using LCA for nanotechnology (CORDIS 2007). The workshop participants agreed that there currently are some serious data limitations, but that the basic methodology is applicable. They also suggested an approach for incorporating life cycle thinking into nanotechnology evaluation. It begins with a traditional check for obvious harm. Next, it conducts a traditional LCA without toxicity study, followed by toxicity- and risk-assessment sets of questions. Finally, the approach combines the LCA and risk analysis (RA) and conducts a scenario analysis. A key theme of this book, risk assessments frameworks that combine LCA and RA are discussed in Chapters 6 and 7.

Life cycle thinking forms the underpinnings for Nano LCRA, the adaptive life cycle risk assessment framework proposed in this book (Chapter 6) and the frameworks described in Chapter 7. This type of approach has broad applicability for industry, governments, and people who are genuinely interested in or concerned about understanding and managing the potential impacts of nanomaterials and nanotechnologies.

Currently, there is a situation of limited data availability for analyzing the impacts of nanotechnology and nanomaterials. Therefore, the tools of LCA, LCIA, and even risk analysis cannot provide quantitative estimates today. However, taking a *screening-level* approach allows these concepts to be incorporated into assessments of new technologies and substances, to improve the ability to identify adverse health and environmental impacts. Since nanotechnology innovation is occurring so quickly, adopting a proactive, science-informed approach that formalizes life cycle thinking into the assessment and management of risk will lead toward a sustainable path for nanotechnology development.

References

Ames, B., and L. S. Gold. 2000. Paracelsus to parascience: The environmental cancer distraction. *Mut. Research* 447:3–13.

ATSDR. 2005. Toxicological profile for lead (draft for public comment). Agency for Toxic Substances and Disease Registry, Centers for Disease Control, Atlanta.

Bossi, R., F. F. Riget, R. Dietz, C. Sonne, P. Fauser, and D. M. Vorkamp. 2005. Preliminary screening of perfluorooctane sulfonate (PFOS) and other fluoro-chemicals in fish, birds and marine mammals from Greenland and the Faroe Islands. *Environmental Poll.* 136(2): 323–29.

Brown, H. S., B. J. Cook, R. Krueger, and J. A. Shatkin. 1997. Reassessing the history of hazardous waste disposal policy in the United States: Problem definition, expert knowledge, and agenda-setting. *Risk: Health, Safety Environment.* 8:249–72.

California Environmental Protection Agency. 1999. Public Health Goal for Methyl Tertiary Butyl Ether (MTBE) in Drinking Water. Available at http://oehha.ca.gov/water/phg/pdf/mtbe_f.pdf

Cassee, F. R., E. C. van Balen, C. Singh, D. Green, H. Muijser, J. Weinstein, and K. Dreher. 2011. Exposure, health and ecological effects review of engineered nanoscale cerium and cerium oxide associated with its use as a fuel additive. *Crit. Rev. Toxicol.* 41(3): 213–29.

Cho, W. S., R. Duffin, C. A. Poland, S. E. Howie, W. MacNee, M. Bradley, I. L. Megson, and K. Donaldson. 2010. Metal oxide nanoparticles induce unique inflammatory footprints in the lung: Important implications for nanoparticle testing. *Environ. Health Perspect.* 118:1699–1706.

CORDIS. 2007. Nanotechnology and the environment: Beauty rather than the beast? http://cordis.europa.eu/fetch?CALLER=NEWSLINK_EN_C&RCN =27711&ACTION=D

Davis, J. M., and W. H. Farland. 2001. The paradoxes of MTBE. *Toxicol. Sci.* 61(2): 211–17.

Energetics. 2012. Case Study: Stagecoach 2007. Update web link with: http://www.energenics.co.uk/case-studies/case-study-stagecoach-2007/

EU Joint Research Centre. 2007. Life cycle thinking. http://lca.jrc.ec.europa.eu/lcainfohub/lcathinking.vm

HEI. 2001. Evaluation of human health risk from cerium added to diesel fuel. Communication 9. Health Effects Institute, Cambridge, MA. http://pubs.healtheffects.org/view.php?id=172

Hess, D. 2007. What is a clean bus? Object conflicts in the greening of urban transit. *Sustainability: Science, Practice, Policy* 3(1): 45–58. http://sspp.proquest.com/archives/vol3iss1/0608-027.hess.html

ICON. 2010. Good nano guide wiki. Beta. International Council on Nanotechnology. http://goodnanoguide.org

Jobling, S., R. Williams, A. Johnson, A. Taylor, A. Gross-Sorokin, M. Nolan, C. R. Tyler, R. van Aerle, E. Santos, and G. Brighty. 2006. Predicted exposures to steroid estrogens in U.K. rivers correlate with widespread sexual disruption in wild fish populations. *Environ. Health Persp.* 114 (Supp. 1): 32–39. http://ehp03.niehs.nih.gov/article/info:doi/10.1289/ehp.8050

Koehler, D. A., and A. D. Hecht. 2006. Sustainability, well being, and environmental protection: Perspectives and recommendations from an Environmental Protection Agency forum. *Sustainability: Science, Practice, Policy* 2(2): 22–28. http://sspp.proquest.com/archives/vol2iss2/0610-032.hecht.html

Kolpin, D. W., E. T. Furlong, M. T. Meyer, E. M. Thurman, S. D. Zaugg, L. B. Barber, and H. T. Buxton. 2002. Pharmaceuticals, hormones, and other organic wastewater contaminants in U.S. streams, 1999–2000: A national reconnaissance. *Environ. Sci. Technol.* 36(6): 1202–11.

McDonough, W., and M. Braungard. 2002. *Cradle to Cradle.* New York: North Point Press.

Mehlmann, M. 2002. Carcinogenicity of methyl-tertiary butyl ether in gasoline. *Annals New York Acad. Sci.* 982: 149–59.

Mennear, J. H., and E. E. McConnell, 1997. Carcinogenicity studies on MTBE: Critical review and interpretation. *Risk Anal.* 17: 673–82.

NIOSH. 2005. Approaches to safe nanotechnology: An information exchange with NIOSH. Draft for public comment. Centers for Disease Control and Prevention, National Institute for Occupational Safety and Health, Atlanta, GA. http://www.cdc.gov/niosh/topics/nanotech/pdfs/Approaches_to_ Safe_Nanotechnology.pdf

———. 2009. Approaches to safe nanotechnology: An information exchange with NIOSH. Department of Health and Human Services. Centers for Disease Control and Prevention, National Institute for Occupational Safety and Health, Atlanta, GA. http://www.cdc.gov/niosh/docs/2009-125/pdfs/2009-125.pdf

NRC. 1996. *Toxicological and Performance Aspects of Oxygenated Motor Vehicle Fuels.* Washington, DC: National Academy Press.

OSHA Act. 1970. General duty clause. Section 5(a) 1. http://www.osha.gov/pls/ oshaweb/owadisp.show_document?p_table=OSHACT&p_id=3359

Park, B., and P. Martin. 2009. Nanotechnology: Applying the 3Rs to risk assessment. National Centre for the Replacement, Refinement and Reduction of Animals in Research. NC3Rs #17. http://www.nc3rs.org.uk/downloaddoc. asp?id=989&page=1074&skin=0

Stern, B. R., and R. G. Tardiff. 1997. Risk characterization of methyl tertiary butyl ether (MTBE) in tap water. *Risk Anal.* 17(6): 727–43.

UN. 1987. Report of the World Commission on environment and development. Annex: Our common future. United Nations, New York.

UNEP. 2001. Life cycle thinking as a solution. United Nations Environment Programme, Paris.

U.S. EPA. 1992. Alternative fuels research strategy. External review draft. EPA 600/ AP-92.002. http://www.epa.gov/ncea/pdfs/mtbe/altfuel.pdf

———. 2006. MTBE. Chap. 14 in Regulatory determinations support document for selected contaminants from the second drinking water contaminant candidate list (CCL 2). EPA report 815-D-06-007. http://www.epa.gov/safewater/ ccl/pdfs/reg_determine2/report_ccl2-reg2_supportdocument_ch14_mtbe. pdf

———. 2007. EPA topics: Lead. http://www.epa.gov/ebtpages/pollmultimedia- polllead.html

———. 2009. Fact sheet: Final third drinking water contaminant candidate list (CCL 3). Office of Water No. 815F09001. September. http://water.epa.gov/ scitech/drinkingwater/dws/ccl/upload/fs_cc3_final.pdf

Winder, C. 1994. History of lead. *LEAD Action News* 2 (2). http://www.lead.org. au/lanv2n2/lanv2n2-17.html

WWF. 2004. European parliamentarians contaminated by 76 chemicals. http://www.panda.org/wwf_news/press_releases/?12622/European-parliamentarians-contaminated-by-76-chemicals

---

Further Reading

Bermudez, E., J. B. Mangum, B. A. Wong, B. Asgharian, P. M. Hext, D. B. Warheit, and J. I. Everitt. 2004. Pulmonary responses of mice, rats, and hamsters to subchronic inhalation of ultrafine titanium dioxide particles. *Toxicol. Sci.* 77(2): 347–57.

Coalition Letter. 2007. Civil society/labor coalition rejects fundamentally flawed Dupont-ED proposed nanotechnology framework. An open letter to the international nanotechnology community at large. 12 April. http://www.etcgroup.org/en/node/610

Davis, J. M. 2007. How to assess the risks of nanotechnology: Learning from past experiences. *J. Nanosci. Nanotechnol.* 7(2): 402–9.

Dupont, E. D. 2007. Nano risk framework (draft). 26 February. http://www.environmentaldefense.org/documents/5989_Nano%20Risk%20Framework-final%20draft-26feb07-pdf.pdf

Hullman, A. 2006. The economic development of nanotechnology: An indicators-based analysis. European Commission, DG Research, Unit: Nano S&T—Convergent Science and Technologies. http://cordis.europa.eu/nanotechnology

Lyon, D., L. K. Adams, J. C. Faulkner, and P. J. J. Alvarez. 2006. Antibacterial activity of fullerene water suspensions: Effects of preparation method and particle size. *Environ. Sci. Technol.* 40(14): 4360–66.

Monica Jr., J. C., M. E. Heintz, and P. T. Lewis. 2007. The perils of pre-emptive regulation. *Nature Nanotechnol.* 2: 68–70.

Oberdörster, G., J. Ferin, and B. E. Lehner. 1994. Correlation between particle size, in vivo particle persistence, and lung injury. *Environ. Health Perspect.* 102(6): 173–79.

Oberdörster, G., E. Oberdörster, and J. Oberdörster, 2005. Nanotoxicology: An emerging discipline evolving from studies of ultrafine particles. *Environ. Health Perspect.* 113(7): 823–39.

Roco, M., B. Harthorn, D. Guston, and P. Shapira. 2010. Innovative and responsible governance of nanotechnology for societal development. In *Nanotechnology research directions to 2020: Retrospective and outlook*, ed. M. C. Roco, C. A. Mirkin, and M. C. Hersam, chap. 13. NSF WTEC report. Berlin and Boston: Springer.

Shvedova, A. A., E. R. Kisin, R. Mercer, A. R. Murray, V. J. Johnson, A. I. Potapovich, Y. Y. Tyurina, et al. 2005. Unusual inflammatory and fibrogenic pulmonary responses to single-walled carbon nanotubes in mice. *Am. J. Physiol.: Lung Cellular Molecular Physiol.* 289: L698–L708.

Starr, C. 1969. Social benefit versus technological risk. *Science* 165(3899): 1232–38.

U.S. EPA. 2006. Regulatory status update: Ion generating washing machines. http://www.epa.gov/oppad001/ion.htm

Warheit, D. B., T. R. Webb, K. L. Reed, S. Frerichs, and C. M. Sayes. 2007. Pulmonary toxicity study in rats with three forms of ultrafine-TiO$_2$ particles: Differential responses related to surface properties. *Toxicology* 230(1): 90–104.

Woodrow Wilson Center for Scholars Project on Emerging Nanotechnology. 2007. A nanotechnology consumer product inventory. http://www.nanotechproject.org/index.php?id = 44&action = view

Xia, T., M. Kovochich, J. Brant, M. Hoke, J. Sempf, T. Oberley, C. Sioutas, J. Yeh, M. R. Wiesner, and A. E. Nel. 2006. Comparison of the abilities of ambient and manufactured nanoparticles to induce cellular toxicity according to an oxidative stress paradigm. *Nano. Lett.* 6(8): 1794–1807.

4

The State of the Science: Human Health, Toxicology, and Nanotechnology Risks

Richard C. Pleus

CONTENTS

4.1 Introduction

4.1.1 What Is the Concern about Nano-Sized Objects?

The last few years have seen a large increase in the use of nanomaterials[*] in consumer and other products. As noted many times, nanosized objects, or nano-objects,[†] and especially their subset, nanoparticles,[‡] will offer tremendous technical opportunities to commerce in nearly every facet of industry. These intentionally engineered particles may solve many problems we currently face in fields as various as energy, fuel, transportation, medicine, building materials, and consumer products. However certain properties of nano-objects suggest that they may pose possible health and environmental risks that ought to be proactively considered. What properties of nanomaterials raise these concerns? They turn out to be the same properties that materials scientists find useful: engineered nano-objects are small and, when compared with bulk materials, have unique physicochemical properties (e.g., surface charge and agglomeration state).[§]

Most of the concerns about the possible health effects of nano-objects are related to size. But what leads scientists to speculate that size matters for human health concerns? The concern comes from years of toxicological research related to exposures from particles, fumes, aerosols, and gases. Most textbooks in toxicology have chapters devoted to respiratory toxicology that discuss how particle size influences the deposition of contaminants in the lungs. Basically, current scientific understanding is that as the size of the particle decreases, it will deposit farther into the lungs, including the alveoli, the air sacs where oxygen and carbon dioxide transfer across single-cell barriers. There are plenty of historical examples of chemicals or fibers that have been small enough to reside deep in the lung

---

[*] A nanomaterial is a material with any external dimension in the nanoscale (between 1-100 nanometers (nm)) or having internal structure or surface structure in the nanoscale (definitions 2.1, 2.16) (ISO 2008).

[†] A nano-object is a material with with one, two, or three external dimensions in the nanoscale (definition 2.2) (ISO 2008).

[‡] A nanoparticle is a nano-object with all three external dimensions at the nanoscale (definition 4.1) (ISO 2008).

[§] A bulk material is defined as a material that has been commonly used and not found to have one dimension at the nanoscale.

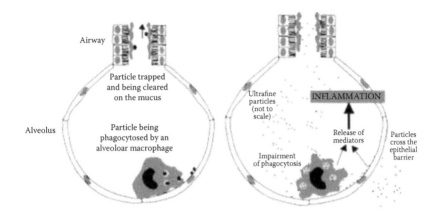

FIGURE 4.1
Diagram of the representation of the hypothetical event after exposure to ultrafine particles. (*Source*: Copyright 2001 BMJ Publishing Group Ltd. Figure reproduced from Donaldson et al. 2001 with permission from BMJ Publishing Group Ltd.)

and are associated with adverse effects. Some examples include coal dust, aerosols from the application of pesticides, asbestos fibers, and air pollutants (such as particulate matter of the ultrafine [UFP],* 2.5-micron [$PM_{2.5}$], or 10-micron size [PM_{10}]). Not all small particles have demonstrated a cause-and-effect relationship, as seen in Figure 4.1.

Given that nano-objects are considered new to toxicology, what scientists logically do to develop their testing hypotheses is to examine the large body of work related to fine or ultrafine particulate matter and extrapolate it to nano-objects. Whether wholesale adoption of this literature will prove to be in harmony with the future body of nanomaterial research, particularly regarding the potential for adverse effects presented by nanoparticles, is likely doubtful. But the current literature on fine or ultrafine particulate matter is a reasonable starting point.

Though particle size is a key issue, we are learning that there may be more important scientific matters that drive the potential for health effects resulting from nanomaterials. Nanoparticle toxicological effects are complex and involve a variety of factors, including physicochemical characteristics, particle-cellular interactions, routes and degrees of exposure, biokinetics, logistics, and other considerations (Warheit 2010). In light of these factors, for instance, a materials scientist may be able to rearrange the atoms of a designed nanomaterial in such a way as to reduce toxicity

---

* Ultrafine particle matter is defined as particles less than 100 nm in diameter. These particles are smaller in size compared to particle matter PM_{10} and $PM_{2.5}$ particle classes, which respectively have diameters of 10 and 2.5 microns. Examples of UFPs include diesel emissions and combustion aerosols (U.S.EPA 2011).

but maintain its engineered purpose. Nanotoxicology is still in the process of incorporating these factors into its assessments. Thus, with diligent application of a sound scientific strategy, research into nanomaterials and the many factors that bear on their toxicological effects will likely bring a new understanding to toxicology that will apply to many compounds we have assessed and will assess in the future.

4.1.2 What Is the Process of Health-Focused Research into a New Technology?

Health-focused research is often scarce for an emerging technology. Scarcity prompts opportunity and action, often quickly for those who are nimble. Who wouldn't want to be the first scientist to discover a new disease? Thus the idea of filling the void is an incentive for conducting research. However, the quality of pioneering studies attempting to fill the void is often poor for many reasons. The speed required to be the first in a new field is one reason. While having a study published early puts a researcher at the forefront of new knowledge as well as the funding cycle, much of the scientific information that needs to be considered can be overlooked. For example, researchers may purchase testing material from a supplier that may not be relevant to the industrial material being used. For nanomaterials, current research suggests that the time a material sits in a container can alter the physicochemical properties of the material, so what is tested today may not be what is tested next month. Another concern is that researchers can overstate results to make a study sound more exciting than it is. Furthermore, toxicologists are trained to test for adverse outcomes. Thus, the information that automatically is the result of toxicity testing is the presence or absence of an adverse effect, irrespective of environmental conditions. For example, what may not be understood is the level of exposure to or dose of a given material needed to cause an adverse effect, which is crucial in determining the risk posed. A given material may need to be applied in doses that are not relevant to environmental concerns before the appearance of an adverse effect.

Funding plays a part in the initial scarcity of health-focused research into new technology as well. The entities that typically fund research, such as governmental agencies or national institutes of health, are often slow to jump on research until the pressure is so great that action is needed. Until that time, research monies are likely bankrolled by existing monies from granting organizations (e.g., money from currently funded non-nano grants). Initial work in a new field thus generally occurs where it intersects with an established field of research. Access to funds to conduct research thoroughly comes later, when researchers provide evidence that proposed studies are feasible and have scientific and public merit.

4.1.3 The Role of Media

The media also play a role in helping or hindering the development of technology. The role of scientists is to conduct properly designed studies, accurately report the results, and then, using standard practices and with sufficient data, evaluate possible health or environmental risk. The role of the media is to report stories and not, unfortunately, to educate. Thus, at times the media fail to present scientific information in the appropriate scientific context; instead, the media frequently direct the readers' attention to a conflict about science (e.g., two scientists who disagree) without grounding that conflict in the science. Readers then base their opinions on perceptions generated by the incomplete media coverage rather than on an understanding of the risks and benefits of technology. This unfortunate outcome can be compounded if scientists fail to accurately report results or accurately evaluate the data. It is thus up to scientists to present information as accurately as possible. This will help the media ground their reports in science and, therefore, help ensure that the public is well informed.

In the last five to six years, three stories were reported on the possible health aspects of nanotechnology. The first is a report from a European paper dated April 14, 2006. The product at issue was a cleaner called "Magic Nano." The news report stated:

> Not much is known about the ingredients in Magic Nano, a household glass and tile sealant, but its recent recall may have been a first for the nanotechnology industry.... Between March 27 and March 30, it said 97 people who reportedly used the aerosol spray claimed to suffer from health problems ranging from trouble breathing to six cases requiring hospital treatment in which water accumulated in the lungs (pulmonary edema). (Needle 2006)

The point of the article—that people were reportedly becoming ill by using a product—was clearly an important health concern that needed to be aggressively addressed. The article implied that nano-objects were a component of the cleaner and could be the cause of the illnesses. However, a subsequent investigation revealed that the product did not contain nano-objects. Unfortunately, it took time for the investigation to reach this conclusion. In the end, the article aroused little public attention; the facts came forward; and the erroneous attribution of risk to nano-objects did not affect the development of nanotechnology.

Another instance involves a case study that was published in the *European Respiratory Journal* as a peer-reviewed article (Song, Li, and Du 2009). Song and colleagues reported that:

> Seven young female workers (aged 18–47 years), exposed to nanoparticles for 5–13 months, all with shortness of breath and pleural effusions were admitted to hospital. Polyacrylate, consisting of nanoparticles, was confirmed in the workplace. Pathological examinations of patients' lung tissue displayed nonspecific pulmonary inflammation, pulmonary fibrosis and foreign-body granulomas of pleura.

Having a study published in a peer-reviewed journal article is a more scientifically valid approach. Publishing in a peer-reviewed journal raises the standard for scientific work, but it is not fail-safe, and the reader is still obligated to evaluate the scientific merits of the work, the experimental design, and the interpretations of the data by the authors. At the time, the public health community believed that if the Song study did describe the first case of nano-related occupational health problems, then it would become an important study. However, after reading the journal article carefully, many scientists became concerned about the conclusions of the authors in light of the lack of exposure data. For example, the National Institute of Occupation Health and Safety (NIOSH) reported the following in their science blog after reviewing the paper:

> There is no doubt about the severity of the heart and lung disease in these workers. One worker died in the hospital 16 days post-surgery and another died of respiratory failure 21 months after symptom onset. The question is, did their exposure to nanoparticles cause or contribute to their disease? Unfortunately, the study cannot scientifically answer that question for us because of a lack of exposure data.
>
> Despite the certainty inherent in their use of the term "nano material-related disease," the study authors have not provided us with scientific evidence nanoparticles were or were not the cause of the tragic medical consequences in these seven workers. (Howard, Geraci, and Schulte 2009)

It is therefore clear that great care must be taken when examining and reporting on results presented in the literature, even when such findings have been published in a peer-reviewed journal.

Researchers' use of key terms can also garner the attention of the media. For example, Poland and colleagues titled their article "Carbon Nanotubes Introduced into the Abdominal Cavity of Mice Show Asbestos-Like Pathogenicity in a Pilot Study" (Poland et al. 2008). This title led to media attention because the public associates asbestos with severe health effects. It is important to note that the Poland article reports the results from a pilot study. Also, in this study, carbon nanotubes (CNTs) were injected abdominally, and administering the material required considerable sample preparation. The results of this type of study need more applicable research in order to place it into proper context.

It is clear that more studies have been conducted since the publication of the first edition of this book. Toxicologists find adverse effects because that is their job. Therefore, as scientists, we need to do a better job of accurately portraying the risks and/or benefits of science to the public and the media.

4.1.4 What Are the Most Common Nano-Objects Used in Commerce Today?

The types of nano-objects that garner the most concern in industrial and commercial application include metal oxides (e.g., titanium dioxide and zinc oxide), CNTs, and quantum dots (e.g., cadmium selenide and zinc sulfide). Using the properties of nano-objects for medications and medical devices is also an important component of introducing this technology into commerce. TiO_2 nanoparticles are mainly used in industrial products, pharmaceuticals and personal care products (e.g., cosmetics), protective coatings, building products, white pigments, plastics, ceramics, cosmetics, solar batteries, and food additives.

While this chapter focuses on CNTs, other nano-objects are also used in commerce. CNTs are carbon allotropes, which are different structural modifications of carbon that form cylindrical structures. Because of their chemistry and structure, they possess unique chemical and physical characteristics. Several forms of CNTs exist and include: single-walled carbon nanotubes (SWCNTs),[*] double-walled carbon nanotubes (DWCNTs),[†] and multiwalled carbon nanotubes (MWCNTs).[‡] The diameters of these structures range from 1 nanometer (nm) (SWCNTs) to 100 nm (MWCNT) (NIOSH 2010). The length of the cylinders varies from 1 nm to many tens of micrometers long (NIOSH 2010). It is not clear at this time what the upper end of the length of these cylinders could be. Several scientists and authoritative bodies have reported that CNTs are physically strong (the elastic modulus of a single CNT wall is about 1000–3000 GPa [gigapascal], making CNTs as stiff as a diamond) (Rogers et al. 2011; Royal Society 2004; Yu et al. 2000). In addition to strength, CNTs have unique electrical and thermal properties. CNTs are also flexible and resilient, allowing them to bend under high stress and spring back to their original shape. CNTs are chemically stable and inert. Because of the strong carbon bonding that holds them together, these tubes do not readily bind with other atoms (Rogers et al. 2011). These properties provide great materials science

---

[*] SWCNTs consist of a single cylindrical graphene layer (ISO 2010).
[†] DWCNTs are composed of only two nested, concentric SWCNTs (ISO 2010).
[‡] MWCNTs are composed of nested, concentric or near-concentric graphene sheets with interlayer distances similar to those of graphite (ISO 2010).

characteristics for use in commerce, but they also define the potential toxicological characteristics as well. For example, these qualities of stiffness and chemical inertness also characterize materials that may not degrade easily in the body or in the environment.

CNTs are used in bicycle components, tennis rackets, golf clubs, and car bumpers. They have also been used in the fuel lines of cars to prevent the buildup of static electricity that might ignite the fuel (Rogers et al. 2011). Silver is used as an antimicrobial for a number of consumer products, including medical devices, food packaging, and clothing.

4.1.5 General Background on Toxicity Testing

The purpose of toxicological experimentation is to assess the potential effects on humans and the environment resulting from exposure to a substance. The toxicity of a substance describes its capacity to cause harm to a living organism, and is generally a result of its hazard (potential harmful effects based on physicochemical properties of the substance) and exposure (quantity of substance the organism comes into contact with; manner of contact). Properly designed experimental studies in toxicology are helpful in reducing uncertainty regarding a substance's toxicity. The intention for all toxicological experiments is to obtain reliable information that includes data related to:

- Dose and response;
- Types and severity of adverse effects;
- Mechanism of action (including upstream biochemistry);
- Any period(s) of time when the organism is particularly sensitive to exposure (e.g., fetal development).

This chapter provides a cursory review of the available studies addressing these dimensions of toxicity for CNTs and makes a high-level assessment of the risk based on these data at this point in time. Note that there is a further concept to introduce with CNTs: They are nano-objects, but they are also considered nanofibers.[*] This provides a level of confusion in the literature that still has not been fully addressed. Thus the interested reader may want to also consider fiber toxicology literature for further information.

The general types of testing include studies conducted *in vitro* or *in vivo*. *In vitro* refers to studies conducted in a "test tube," meaning that tests are conducted on isolates of animal tissues, such as cells or parts of cells, in small containers such as a petri dish. These provide information on

---

[*] A nanofiber is a nano-object with two similar external dimensions in the nanoscale, with the third dimension significantly larger (definition 2.5) (ISO 2010).

nonorgan toxicology, meaning that these tend to be biochemical reactions that occur potentially in many types of cells. These studies are useful, as they generally provide information on both the chemistry of the material and its interaction with the biological material.

In vivo refers to studies conducted in whole-animal systems. These provide information on organ toxicology, meaning that these types of studies tend to determine effects on an organism's organ systems. Animals used in *in vivo* studies can range across the whole animal kingdom, but for human health issues, *in vivo* mainly refers to studies performed on laboratory animals (e.g., rodents, canines, felines, primates) or on humans. Other types of scientific studies that are often classified as *in vivo* include human clinical studies and epidemiologic studies. *In vivo* studies are useful in that they provide information on biochemistry as well as the response or adaptation of the animal to the toxicant, including assessing protective mechanisms. Well-conducted experimental studies assume the use of the scientific method. For example, the design of experiments that test the toxicity of a material must be such that the experiments are repeatable and the results are reproducible.

Given the current global interest in limiting the use of animal studies, there is a more vigorous effort to enhance our use of *in vitro* studies. Whether these studies can effectively replace *in vivo* studies in the future seems unlikely; however, it is important to continue to strive to improve *in vitro* studies to be as relevant as possible.

Another area of interest in toxicological testing that has relevance for evaluating nano-objects is *in silico* testing. *In silico* testing involves the use of computers to predict the toxicological outcome of the chemicals or nano-objects in question. While the technique is new, many researchers from many fields are using its methods to predict chemical interactions with data from studies already performed. We expect to see this field grow over the next several decades.

In their efforts to advance the study of toxicology, toxicologists and other medical scientists have developed and adopted procedures for assessing the possible risk(s) of harmful effects of materials and, conversely, the degree of safety. Following these procedures, scientists in government, industry, and academia can make these toxicological risk assessments for human health as well as for the environment. As described in Chapter 2 (and in greater detail in the U.S. National Research Council publication *Risk Assessment in the Federal Government: Managing the Process*), the risk assessment process comprises four steps. These are: (1) hazard identification, (2) dose–response assessment, (3) exposure assessment, and (4) risk characterization (NRC 1983). Toxicological testing, predominantly *in vivo* studies, provides fundamental data for hazard identification, dose–response assessment, and exposure assessment. Risk assessment data are used to derive other information such as worker or residential exposure limits, recommendations for personal protective equipment, and hazard communication documents.

4.1.6 General Information on Toxicology Testing and CNTs

At this time in the evolution of nanotechnology, most of the health-related questions under consideration come from concerns about workers who are involved in the manufacture, production, or use of nano-objects (NIOSH 2010). The working environments of these fields have the greatest concentration of nanomaterial and are also where exposures occur most frequently; thus workers in such environments face the highest risk of health effects. The most significant exposure route for workers is inhalation, rather than ingestion or dermal. Possible exposures from materials released in the environment, such as emissions from factories, are only slightly less important than occupational exposures from an exposure-assessment standpoint. Following that, exposures to consumers of products with CNTs may also be a significant risk.

Acute toxicity studies are conducted at high concentrations for short periods of time, primarily to determine the potency of a chemical. A common numerical value from this type of study is called the Lethal Concentration of half the study population (LC_{50}; for oral dose, it is called LD_{50}, the Lethal Dose of 50% of the population). For subchronic testing, a longer period of time is common; exposures are generally five days per week, six hours per day, for ninety days. Subchronic toxicology testing provides information about how an organ or organ system will be affected. Hematological, clinical blood chemistry, and urine analyses are obtained during the exposures. Following exposure, animals are sacrificed, and complete pathological and histological evaluations are conducted. By comparison, chronic studies are conducted to determine the effects of long-term exposures at levels where acute toxicity is not apparent. Daily exposures can be similar to those in subchronic tests except that the duration is up to two years. In some cases, exposures are conducted twenty-four hours a day for seven days a week. Studies were retrieved for acute and subchronic testing of CNTs, but not for chronic testing of CNTs.

The costs for conducting toxicity studies are high. As of 2011, the cost of subchronic inhalation toxicity testing for a CNT was approximately $750,000. Costs for characterizing physicochemical parameters are added to that amount. Under the Toxic Substances Control Act (TSCA), U.S.EPA is currently required to assess and regulate risks to human health and the environment before a new chemical substance is introduced into commerce (U.S.EPA 1997). These costs can be barriers to commerce for small companies in the United States or those who would like to export materials to the United States. One way to address both the demand to have testing conducted and the cost to do so is to form consortia. In 2010, the

NanoSafety Consortium for Carbon (NCC) was formed with fifteen small-to-medium-size companies to meet the requirements set by U.S.EPA (NCC 2010; U.S.EPA 1976, 1997). By working together, risks from nano-objects can be assessed and costs shared.

Thus far, the studies that have been published concerning occupational exposures focus largely on the possible concentrations of engineered nano-objects found in the workplace. No information as of this date has reported health outcomes from exposures to CNTs (e.g., workers in these facilities). There will likely be studies to evaluate health effects in the future. The best assessment so far is that CNTs are unlikely to cause an acute health effect, and that any possible health effects will be a result of longer-term exposures at lower doses. There has not been enough time in the history of the use of these materials for the latter situation to have occurred. Such long-term health-effect data are also limited because many companies have worked diligently to reduce long-term exposure risk by installing state-of-the-art ventilation systems and requiring workers to use personal protective equipment even before studies have been conducted.

Other exposures are likely to occur, which will become important for toxicology research. There are some studies that report discarded items in waste streams, landfill leaching, and waste incineration products (with the exception of high-temperature incineration) (Köhler et al. 2007). Thus, occurrence information will be useful. No quantitative data on consumer exposure to CNTs were identified.

Toxicity testing for inhalation exposures is not a trivial exercise, as there are many variables to control. Generation of a test atmosphere and contaminant, mixing of the test material and atmosphere, and movement of the atmosphere through the experimental system for treated and control animals are elements of toxicity testing for inhalation that are difficult to perform no matter what substance is being tested. Toxicity testing for inhalation of CNTs and other nano-objects is even more challenging because CNTs tend to agglomerate,* which makes generation of homogeneous atmospheres of CNTs difficult. Measurement and characterization of the materials that animals have been exposed to, as well as the cleanup and disposal of the testing materials, must also be considered in experimental design.

Study guidelines are available for some inhalation studies. The U.S.EPA Office of Prevention, Pesticides, and Toxic Substances (OPPTS) has guidelines for acute and subchronic inhalation toxicity testing (870.1300 and 870.3465, respectively) (U.S.EPA 1998). The Organization for Economic Cooperation and Development (OECD) also has developed guidelines for

---

* An agglomerate is a collection of weakly bound particles or aggregates or mixtures of the two where the resulting external surface area is similar to the sum of the surface areas of the individual components (definition 3.2) (ISO 2008).

animal testing. A number of issues related to nanomaterial testing raise important experimental questions of the testing guidelines, including the tendency of some nano-objects to bind with test reagents or to aggregate* in test media. The researcher is cautioned to be alert to the possible influences of nano-objects and nanoparticles when designing and conducting toxicological research; however, the initial consensus opinion is that testing guidelines are adequate (Environmental Defense and DuPont Nano Partnership 2007; OECD 2010).

For inhalation studies, there are three categories of exposures that are considered. They are: nose only, head, and whole-body exposures. However, there are some adaptations to these types of studies that have been reported in CNT studies. They include delivery into the lungs by intratracheal instillation or pharyngeal aspiration, which are ways of delivering a dose to the lung via the upper airways. While inhalation is considered the most physiologically relevant exposure pathway because it provides a natural route of entry into the lungs and addresses important factors that may influence dose to the subject, instillation and pharyngeal aspiration have been more useful, as the quantitation of dose is better controlled for these methods as opposed to inhalation exposures.

One of the greatest challenges in laboratory testing, which then has a direct consequence on the reliability of the data, is the variation across experimental subjects and the delivery of a known quantity of material. In the case of inhalation studies, test animals all have in common an airway, including the trachea, bronchi, bronchioles, and alveoli. However, the respiratory system of the test subject varies anatomically across rodents, canines, felines, and primates. Each test animal's respiratory system will have different patterns of deposition of small particles, which leads to difficult interpretations when extrapolated to the human.

The second issue that has a direct effect on the reliability of the data is that the generation of consistent levels of nano-objects in air is difficult. For example, MWCNTs tend to aggregate and form agglomerates. When that happens, the particles increase in size and mass, and consequently the deposition of test material may be different than expected. If not controlled in experimentation, the dose can be greatly affected, significantly skewing the results or even making the results unreliable. Similarly, agglomeration and aggregation may have effects in *in vivo* studies as well, including inhalation studies. Thus, it is important to assess if agglomeration occurs when assessing CNTs, so as to better understand the possible influence agglomeration may have on the results of inhalation studies.

---

* An aggregate is a particle comprising strongly bonded or fused particles where the resulting external surface area may be significantly smaller than the sum of calculated surface areas of the individual components (definition 3.3) (ISO 2008).

Intraperitoneal/intravenous (i.p./i.v.) injection and dermal exposure methods are other ways used to deliver CNTs, but these may or may not be an appropriate exposure route, depending on the scientific question being raised. For example, in the studies by Poland et al. (2008) and Takagi et al. (2008), CNTs were administered intraperitoneally. Intraperitoneal injection means that the exposure was given by a syringe into the abdominal cavity of the animal. Validation of this type of experiment will need to be conducted. Intraperitoneal administration bypasses many protective mechanisms of the body. For example, materials that are ingested are exposed to gastric juices; absorbed, depending on the gut contents and physicochemical properties of the material; and, if absorbed, pass through the liver, where metabolism (e.g., detoxification) can take place. However, some believe intraperitoneal administration provides screening information on effects of fibers, such as asbestos. Exposure via inhalation or dermal contact is expected during the manufacturing and handling process; thus, toxicity studies should focus on these exposure pathways.

Thus far, the only information on possible human health effects from CNTs is from either *in vitro* or *in vivo* animal studies. No other types of human studies, such as epidemiological studies, are reported at the time of writing. NIOSH has plans to conduct an epidemiological study, but the dates on which it will begin and end are not yet clear.

4.2 What Research Has Determined Thus Far

4.2.1 What Do We Know about the Nano-Object?

4.2.1.1 Physicochemical Characteristics of Nano-Objects

What we mean by a physicochemical characterization of nano-objects is simply stated as information regarding what a given material looks like, what it is made of, and what special properties result from the engineering of this nano-object. Of course, nano-objects cannot be characterized without special laboratory equipment. Examples of equipment used for physicochemical characterization are transmission electron microscopy (TEM), atomic force microscopy (AFM), field-flow fractionation (FFF), energy-dispersive X-ray (EDX) spectroscopy, and inductively coupled plasma mass spectroscopy (ICP-MS). These are expensive instruments that require skilled technicians to operate. Scientists hope that physicochemical characterization will allow scientists to predict the toxicity of an engineered nano-object, possibly even before the nano-object is tested in animals.

The list of eight parameters and their descriptors identified by the International Organization for Standardization (ISO) as important in toxicological assessments is shown in Table 4.1. This list was assembled by an international group of health scientists, toxicologists, and metrologists (ISO 2012). For addressing what a manufactured nano-object looks like, the following parameters are relevant: particle size/distribution, aggregation/agglomeration state in relevant media, shape, and surface area/specific surface area. For addressing the question of what manufactured nano-objects are made of, the following parameters are relevant: composition, purity, and surface chemistry. For addressing the unique properties of manufactured nano-objects, the following parameters are relevant: surface charge, solubility, and dispersibility.

After reviewing over 100 different CNT studies published in the last ten years, there are only a limited number of *in vitro* and *in vivo* studies that characterize any of the eight physicochemical parameters identified by the ISO. No study assessed all eight parameters. Across all studies reviewed, the most common parameters assessed include a reporting of shape, size, aggregation/agglomeration, and solubility/dispersibility. To a lesser degree, surface area and surface chemistry were reported. Another observation regarding these studies is that the methods used to characterize a given parameter were not provided in sufficient detail in the published manuscripts. The lack of physicochemical characterization makes it difficult to rely on the data and draw conclusions about the toxicity of CNTs. Toxicology studies need to both characterize the physicochemical parameters and provide sufficient detail. Other researchers may then use the obtained data to properly compare and contrast experimental results, and scientists in the future will be able to better understand a particular result reported today.

Before the literature is discussed, it is important to note that the physicochemical characterization of a nano-object provided by a supplier's commercial data is not sufficient for toxicological testing purposes. First, the characterization is likely to be for industrial or commercial applications rather than for toxicology research. Second, there are instances where characterization of a material at one time is different from the characterization obtained at another time, which demonstrates that some nanomaterials age or react with the environment. Third, when testing a material, toxicologists are likely more interested in what happens when the test material is prepared and in the characterization of the material after it is delivered to the test subject. The concern is that preparing the test material for administration will change the material's physicochemical characterization. Thus the prepared test material that toxicologists examine may have significant differences in toxicity and mechanism from the supplied raw material.

Now the findings from the literature are reported. Some studies support the concept that surface area is the dose measurement that best predicts lung toxicity. As an example, in a study by Oberdörster, Oberdörster, and Oberdörster (2005), rats and mice were exposed to two different sizes

TABLE 4.1

List of Physicochemical Characterizations for Manufactured Nano-Objects
Submitted for Toxicological Testing

Parameter	Descriptor
Aggregate/agglomeration state	*Aggregate*: Strongly bonded or fused particles where the resulting external specific surface area might be significantly smaller than the sum of calculated specific surface areas of individual components *Agglomerate*: Collection of weakly or loosely bound particles or aggregates or mixtures of the two in which the resulting external specific surface area is similar to the sum of the specific surface areas of the individual components
Shape	A geometric description of the extremities of the nano-objects or collection of nano-objects, aggregates, and agglomerates that make up the material under investigation
Surface area/mass-specific surface area/volume-specific surface area	Surface area is the quantity of accessible surface of a powdered sample when exposed to either gaseous or liquid adsorbate phase Surface area is conventionally expressed as a mass-specific surface area or as a volume-specific surface area where the total areal quantity has been normalized either to the sample's mass or to a volume
Composition	Chemical information and crystal structure of the entire sample of nano-objects, including: (a) composition and (b) presence or absence of crystalline structure, including lattice parameters and space group, and impurities, if any
Surface charge	Electric charge on a surface in contact with a continuous phase
Surface chemistry	Chemical nature, including composition, of the outermost layers of the nano-object
Particle size distribution	The physical dimensions of a particle determined by specified measurement conditions
Solubility/dispersibility	*Solubility*: The degree to which a material (the solute) can be dissolved in another material (the solvent) so that a single, homogeneous, temporally stable phase (a suspension down to the molecular level) results *Dispersibility*: The degree to which a particulate material (the dispersed phase) can be uniformly distributed in another material (the dispersing medium or continuous phase)

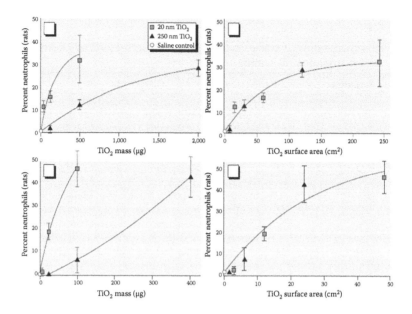

FIGURE 4.2

Percentage of neutrophils in lung lavage of rats (A,B) and mice (C,D) as indicators of inflammation 24 hours after intratracheal instillation of different mass doses of 20-nm and 250-nm TiO$_2$ particles in rats and mice. (A,C) The steeper dose response of nanosized TiO$_2$ is obvious when the dose is expressed as mass. (B,D) The same dose–response relationship as in (A,C), but with dose expressed as particle surface area; this indicates that particle surface area seems to be a more appropriate dose metric for comparing effects of different-sized particles, provided that they are of the same chemical structure (anatase TiO$_2$ in this case). Data show mean ±SD. (*Source*: Copyright 2005 by Oberdörster, Maynard, et al.; licensee BioMed Central. Figure reproduced from Oberdörster, Maynard, et al. 2005 with permission from BioMed Central.)

of titanium dioxide (TiO$_2$) particles—20 and 250 nm in diameter—via tracheal instillation. The results, which were replicated in both species, demonstrate that TiO$_2$ (20 nm) had a much greater percentage of neutrophil (immune cell) response in lung lavage than did TiO$_2$ (250 nm) when both types of particles are instilled at the same mass dose. However, when the instilled doses are matched to surface area, the neutrophil response fit the same dose–response curve. This demonstrates that particle surface area for particles of different sizes but of the same chemistry, such as TiO$_2$, is a better dose metric than is particle mass or particle number (see Figure 4.2).

Another example of how physicochemical properties can influence toxicity results is seen in work overseen by the National Cancer Institute (NCI), which conducts preclinical characterization of nanoparticles for cancer drugs. Dr. Scott McNeil, the director of the laboratory, stated in 2009 that "slight changes to a nanoparticle's size or surface chemistry, for

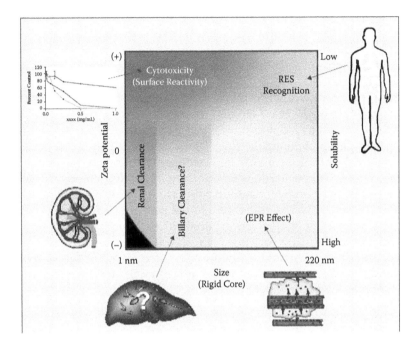

FIGURE 4.3
The physicochemical characteristics of a nanoparticle influence biocompatibility. The figure qualitatively shows trends in relationships between the independent variables of particle size (neglecting contributions from attached coatings and biologics), particle zeta potential (surface charge), and solubility with the dependent variable of biocompatibility—which includes the route of uptake and clearance (shown in green), cytotoxicity (red), and RES recognition (blue). (*Source*: Copyright 2009 John Wiley & Sons, Inc. Figure reproduced from McNeil 2009 with permission from John Wiley & Sons, Inc.)

instance, can dramatically influence a physiological response." As can be seen from Figure 4.3, which is adapted from his 2009 paper, while not producing final results at this early stage, he and his team's assessment of chemotherapeutic results see trends in biocompatibility based on size, surface charge, and hydrophobicity (McNeil 2009).

Another related aspect to the physicochemical characterization of a manufactured nano-object is its purity or impurity. While these terms appear to be opposites of one another, they are not. Purity is the amount of declared substance in the material. This is important, as some research reports the use of highly pure nano-objects of 97%. However, this may not be sufficient information, as the other 3% of the material may contain toxicologically active substances. Impurity describes an unintended constituent present in a manufactured nano-object. The production of manufactured nano-objects may include impurities, such as heavy metals (e.g., iron, nickel, cadmium, etc.). From a toxicology point of view, impurities are relevant if they are present in the material and have toxicological and/ or ecotoxicological importance. Some studies demonstrate that purified and/or modified CNTs appear to be less toxic than nonpurified forms (Jia et al. 2005; Bottini et al. 2006; Dumortier et al. 2006; Murray et al. 2009; Patlolla et al. 2010).

The terms used to describe the purity or impurity of a material include "raw" and "purified." For example, the degree of purity of a SWCNT for a metal is reported as follows: raw "25 wt% Fe" and purified "<4 wt% Fe." The ranges of metal impurities fluctuate quite extensively, from as high as 30% to less than 0.1% for SWCNTs. The degree of impurity, particularly of metals, appears to be an important variable regarding the toxicity found by the study, as a number of authors report that their experimental results differed whether the materials were in the raw or purified form (Lam et al. 2004; Wick et al. 2007; Pacurari et al. 2008). Metal catalysts are used in some manufacturing processes of CNTs. If the metal content, form of metal, and the particular metal continue to affect the outcome of toxicity experiments, then the manufacturing process will be an important consideration in the evaluation of toxicity.

When CNTs are purified, another possible result of the process is that some of the physicochemical properties change (e.g., surface chemistry). Some investigators demonstrate that functionalized (e.g., positively charged or oxidized) CNTs may be more toxic than nonfunctionalized forms (Bottini et al. 2006; Patlolla et al. 2010). Johnston and colleagues (2005) confirm the observation that an acidic and oxidizing treatment strongly modifies the surface chemistry and the agglomeration behavior of CNTs.

The relevance of testing a raw or purified material is arguably related to the question being asked. If the question is about a nano-object's toxicity, studies examining both raw and purified forms are useful. From an environmental perspective, testing material in its raw or "as manufactured" form is likely to provide the most relevant information, as this would be the form a worker is exposed to.

Thus, scientists have underscored the importance of properly characterizing a nano-object's physicochemical properties.

4.2.1.2 How Does the Nano-Object Work?

As with pharmaceuticals and some environmental agents, the more scientists understand the mechanism of action (MOA) of a given compound, the more valuable is the information scientists have for preventing adverse effects; developing protective approaches to using or developing the compound; or engineering out, if possible, the characteristic that causes toxicity. Using the MOA of an agent is the future of toxicology and is crucial to understanding how the physicochemical characteristics of a nano-object are linked to toxicity. The MOA of an agent generally involves a detailed understanding and description of the molecular basis for an effect (adverse or not) so that dose and response can be established in molecular terms. The mode of action (MOa) of an agent is similar, but focuses more on a sequence of key events, such as cytological (cellular) and biochemical (changes in proteins and other molecules) events that lead to an adverse effect.

As an example of an MOA from outside the nanomaterial field, consider the chemical perchlorate, which has been used as a pharmaceutical agent and is an environmental contaminant. Its MOA is well understood. At sufficient dose (e.g., therapeutic levels), perchlorate inhibits the uptake of iodine to the thyroid gland. Inhibition of iodide uptake in itself is not an adverse effect and occurs daily as a response to chemicals that are naturally a part of the food we eat; however, it is the first measurable event on the continuum of effects. If the inhibition of iodide uptake continues for weeks or months and the body's adaptive mechanisms are overcome, thyroid hormones could be reduced. Reduced thyroid hormone, or hypothyroidism, would be the first adverse effect on the continuum.

One area of MOa research is the production of reactive oxygen species (ROS). ROS generation is not specifically tied to a response from nano-objects. ROS generation causes oxidative stress, an imbalance within cells where there is increased reactive oxygen and decreased antioxidants. Reactive oxygen species include superoxide (O_2^-) anion and the formation of hydroxyl ($\cdot OH$) radicals. These products react with proteins and lipids, affecting their functions in enzyme systems, transporters, and cellular structural components. Hydroxyl radicals are thought to be the primary mediators of deoxyribonucleic acid (DNA) damage.

Several *in vivo* and *in vitro* studies have demonstrated increases in ROS formation when exposed to nano-objects. With regard to CNTs, both SWCNTs and MWCNTs have been shown to induce ROS formation in different cells lines (Shvedova et al. 2003; Manna et al. 2005; Sharma et al. 2007; Jacobsen et al. 2008; Pacurari et al. 2008).

What scientists learn from MOA and MOa is information that can be used to develop safe exposure levels to prevent the procession to adverse effects. This is useful information that will provide toxicologists with greater certainty on safe levels.

4.2.2 Toxicity Testing of CNTs

4.2.2.1 General Comments on the Literature

At this point in time, there is a limited amount of literature in every one of the categories examined in this chapter regarding toxicity testing of CNTs, though research has progressed since the writing of the first edition of this book. The works examined in this chapter have general limitations. First, not all published studies are included. The studies herein, for the most part, demonstrate experimental approaches that are clear and adequately described. A very few studies are included, as they represented the newest relevant information.

A second issue is that the reader of the chapter may infer that all CNTs have the same effects. That is not true. SWCNTs and MWCNTs are different in many ways other than whether they are single-walled or multiwalled. While they are often combined in the sections of this text, the reader should not make broad conclusions at this time. The impression one should leave with is that we know much more than five or ten years ago, but we are still early in the toxicology research. This emphasizes the importance of characterizing physicochemical properties of nano-objects so that we may begin to identify those characteristics that lead to toxicity and those that do not.

A third issue is that the effects noted for nano-objects are not applicable when nano-objects are embedded, melded, or incorporated into a matrix, such as a plastic or a composite. Further testing is required to determine what, if any, effects result from such materials.

The reviewed body of literature has some further limitations, including:

1. A general lack of studies for some areas of traditional toxicology research;
2. For some studies, a lack of proper explanation of experimental design (e.g., use of a single dose, lack of a control group); and
3. Inconsistent and insufficient characterization of the nano-object.

Another important issue for these experiments is that, in some cases, not only is just one dose tested, but in some experiments, doses can be rather high and may not be representative of exposures to the general population or to workers handling the materials. What might be reasonable to conclude from these experiments is that the literature is beginning to develop a toxicological database for information on exposure and hazard assessment of CNTs.

The future will likely see more data from dose–response studies that can be better used for human and environmental health risk assessment. Risk assessments can certainly be conducted with the information reviewed thus far, but the scientific community would need to apply a number of conservative factors to ensure that we are not underestimating health risk (see Chapter 8 for NIOSH risk assessment).

4.2.2.2 Absorption, Distribution, Metabolism, and Excretion of CNTs

4.2.2.2.1 What Happens to the Agent When It Enters the Body?

When a person is exposed to a chemical, toxicity is dependent on the chemical entering the body, being metabolized, and then excreted. The process of a chemical entering the body is called *absorption*. Common routes of absorption include ingestion, inhalation, and dermal penetration. Other routes, such as intraperitoneal, intravenous, and intra-arterial, are more commonly associated with medicines than with environmental agents. Once in the body, chemicals are often distributed, meaning that they are transported between locations in the body. Chemicals that enter the body are often metabolized. Metabolism generally makes a chemical less toxic; however, there are a number of cases where metabolism makes a chemical more toxic before it detoxifies. The primary organ of detoxification is the liver. Finally, most chemicals are released from the body by the process called excretion. The most common excretion path is in urine, via the kidneys. Other pathways of excretion are known and include intestinal release, release from the lungs, and release from the skin. Taken together, the absorption, distribution, metabolism, and excretion (ADME) information of a chemical describes how it moves through the body; it is thus important to understand the ADME properties of nano-objects.

4.2.2.2.2 General Observations

At this time, the data suggest that CNTs enter the body primarily from inhalation and secondarily from ingestion as it relates to environmental health and safety concerns. Other routes can be used for CNTs in medical applications. Absorption appears to occur to some extent depending on the route, but how and under what conditions is not understood. Once in the body, CNTs appear not to be metabolized. This will likely be an important area of future research as questions of toxicity are also related to the biochemistry of detoxification of materials. For CNTs that are ingested, CNTs appear not to be absorbed and, hence, are released from the intestines. This will likely be of some interest to the water industry, as CNTs could likely enter the environment, and thus wastewater treatment facilities may need to consider their removal. For CNTs taken into the lungs or via the skin, it is not clear at this time how these materials can be or are excreted. Presently, there is no model linking physicochemical parameters of CNTs to their ADME properties. Since the ADME properties of CNTs are highly dependent on the physicochemical parameters of particular CNT formulations, this might also be an important area of research.

4.2.2.3 Absorption

Subsequent to pulmonary exposure, several studies demonstrate that a considerable fraction of CNTs (often MWCNTs) remain within the lung for up to several months following exposure (Muller et al. 2005; Shvedova et al. 2007; Aiso et al. 2010; Pauluhn 2010; Porter et al. 2010). This provides evidence that these CNTs are biopersistent in the lung. Biopersistence is thought to be derived from the particles' composition and structure. For

MWCNTs following intratracheal administration, about a quarter of the administered dose remains within the lung. Some studies demonstrate that MWCNTs are not efficient in crossing the pulmonary barrier yet were still present in the lungs up to six months later after a single instillation.

4.2.2.3.1 Distribution

The distribution of CNTs to various organs is reported following intravenous exposure, with predominant localization within the liver, lungs, and spleen. *In vitro*, functionalized CNTs are reported to effectively cross biological barriers such as the plasma membrane. Following intraperitoneal administration, SWCNTs are reported to accumulate predominantly within spleen, liver, and urine. They are reported to be excreted via urine following an intravenous injection. MWCNTs are phagocytized by Kupffer cells within the liver, with no apparent toxicity associated with exposure (Cherukuri et al. 2006; Singh et al. 2006).

In the lung, movement of CNTs appears to be related to their type and size. Much more research is needed to elucidate more specifically which types are more or less able to be removed from the lungs. The two general observations thus far are the removal of CNTs from lungs and the migration of CNTs to other locations.

Clearance of CNTs from the lung is reported to be due to the action of macrophages. Macrophage-mediated clearance of CNTs, via phagocytosis, from the lung was demonstrated followed by the death of the macrophage (Porter et al. 2010).

Migration of MWCNTs for the size ranging from 0.5 to 50 µm is noted. Single inhalation exposure has shown that MWCNTs migrated to the subpleura in mice from what were considered high atmospheric concentrations (30 mg/m³), but this was not observed at lower concentrations (1 mg/m³) (Ryman-Rasmussen et al. 2009). MWCNTs of approximately 4 µm in length reached the pleura and induced pleural inflammation fifty-six days after a single aspiration of 10–80 µg in mice. The pleural cavity is the location in which pathogenic fibers are known to elicit diseases such as mesothelioma (Porter et al. 2010).

Thus, across several studies, translocation of CNTs, after inhalation, from the lung to other organs is sometimes reported and is sometimes not observed. For those organs that are reported as damaged without evidence of CNTs, the speculation is that the introduction of CNTs into the body of some test subjects might release and circulate mediating factors rather than the CNTs themselves. More research will need to be conducted to repeat the findings of these studies.

4.2.2.3.2 Metabolism

While it is clear that shorter fibers are cleared more easily by macrophages and that modification of physicochemical characteristics can influence biopersistence, from the studies thus far, it appears that degradation of CNTs in the body may be difficult due to their biopersistent nature. The most recent evidence is that some CNTs can be degraded, although more work needs to be done in this important area. At the Society of Toxicology Meeting in 2012, a research team reported that SWCNTs were susceptible to enzymatic degradation by some enzyme systems using *in vitro* test systems (e.g., eosinophil peroxidase) (Kapralov et al. 2011). Others have shown CNT susceptibility to myeloperoxidases (a human enzyme) expressed in neutrophils (Kagan et al. 2010). This will be a critical area of research in the future. Chemicals that persist in the body and the environment, even if not particularly toxic, generally draw the attention of the public.

4.2.2.3.3 Excretion

After oral administration of MWCNTs in the mouse, the majority of MWCNTs are found in the feces. Some remain within the stomach and small and large intestines, with no detectable transport into the blood. Examination of MWCNTs shows that the nano-objects remained unchanged. This also supports the idea that some CNTs are biopersistent if not excreted. In what might be an immunological reaction to SWCNTs by the test subject, Folkmann and colleagues (2009) report that oral administration and absorption of SWCNTs caused oxidative damage in a dose-dependent manner to the liver and lungs. Interestingly, no SWCNTs were detected in these organs.

4.2.3 Nonorgan Toxicity

4.2.3.1 General Observations

Inflammation and fibrosis appear to be common end points in toxicological experiments in laboratory animals. In the acute phase, inflammation is characterized by formation of polymorphonuclear leukocytes, macrophage influx, and release of proinflammatory cytokines and chemokines. Largely immune reactions, the MOas involving inflammation include cytokines, tumor necrosis factors, chemokines, migration inhibition factors, and production of ROS. When cells are inflamed or damaged, they can release these chemicals, which are important protective defenses against tissue injury or inflammation.

Cytokines are soluble proteins secreted by macrophages. Once present in the site of infection, they can recruit other cells of the immune system, such as neutrophils, eosinophils, or basophils, into the infected area. Neutrophils are specialized to phagocytize and kill intruders (e.g., pathogens, foreign bodies). They can act as soon as an infection arises, and are the principal phagocytic cell in infected and inflamed tissues. Neutrophils have a short life (4 to 10 hours) and die at the site of infection immediately after attacking a foreign particle. The chemicals released by these cells induce inflammation at the site of infection. Cytokines work as signaling mediators, serving as initial triggers of antibody production. Some tumor necrosis factors (e.g., TNF-α) have been reported in many *in vivo* and *in vitro* CNT studies as an indicator of immunological activity, inflammation, and fibrosis. As an example, Chou and colleagues (2008) propose the molecular characterization of SWCNT-induced cytotoxicity in macrophages. In the samples they used, uptake of SWCNTs into macrophages induced oxidative stress in mitochondria and activated nuclear factors and transcription factors (such as NF-*k*B and AP-1, respectively), which led to large-scale production of cytokines and chemokines. Uptake of SWCNTs also induced the expression of protective and antiapoptotic genes and led to the activation of certain white bloods cells called T or T-lymphocyte cells.

4.2.3.2 In Vitro Testing

A large number of *in vitro* studies have demonstrated that both SWCNTs and MWCNTs can induce cytotoxicity in different cell types, including lung epithelial/fibroblast cell lines, human epidermal keratinocytes, human skin fibroblasts, human embryonic lung cell lines, human acute monocytic leukemia cell lines, primary human T-lymphocytes, alveolar macrophages, rat macrophages, and normal and malignant mesothelial cell lines. The cytotoxicity response to CNTs may differ, depending on the experimental conditions (e.g., cell line).

Some of the results of *in vitro* testing demonstrate that responses are only seen, or at least are a more robust response, at times when either SWCNTs or MWCNTs are the test material. This appears to be due to the type of cell line, whether the nano-object is raw or impure, or whether there is agglomeration or aggregation, among other parameters. A number of studies evaluate the role of metals in CNT-induced toxicity, in which investigators quantify for the presence of contaminants or impurities in CNTs. The studies examine metals, including amorphous carbon, iron-rich oxides, and other trace metals (e.g., nickel, cobalt, yttrium, molybdenum). The studies conclude that toxicity is dependent on impurities present on the surface of the CNT. For example, one team of researchers reports that iron-rich SWCNTs are more capable of stimulating oxidative stress in macrophages than their iron-depleted counterparts (Kagan et al. 2006).

Many studies report that SWCNTs can induce toxicity based on dose and time of exposure; however, there are few experiments that adequately assess these fundamental toxicological parameters. The types of cellular responses reported in cytotoxicity tests include a number of different biochemical reactions (e.g., ROS-induced DNA damage), a decrease in cell number, cell apoptosis, and impaired phagocytosis of some cells (e.g., macrophages).

More and more *in vitro* studies have been conducted in the past five years. There is a clear goal by authoritative bodies to move more research to this area and away from conducting *in vivo* studies (e.g., Tox21 Program[*]). The studies conducted thus far are investigating biochemical pathways and data suggest that CNTs cause cytotoxicity. Regarding ROS production, this mode of action appears to be discussed most frequently for cellular toxicity. The dose required for these reactions is not easily understood from the perspective of what type of dose would be required *in vivo*. Although it is desirable to replace *in vivo* studies with *in vitro* studies, the latter are unlikely to provide the information we obtain from *in vivo* studies.

4.2.3.3 Mutagenicity and Genotoxicity

Genotoxic responses are associated with DNA damage and mutations, which may arise through direct or indirect mechanisms on cells, that compromise the survival and function of the cells. Genotoxic responses may be primary (due to some physical chemical characteristics that influence cytotoxicity or genotoxicity) or secondary (due to excessive biopersistent presence of nano-objects capable of inducing inflammation) in nature.

Several *in vitro* and in *in vivo* genotoxicity tests, using different models, have been performed with different types of SWCNTs and MWCNTs. The types of *in vitro* testing reported in the literature include the Ames[†] assay (Kisin et al. 2007), with or without metabolic activation; the Comet[‡] assay (Pacurari et al. 2008); the micronucleus[§] assay (Kisin et al. 2011; Kato et al. 2012); and the chromosome aberration[¶] assay (Asakura et al. 2010). The results of these toxicity tests indicate that for those nano-objects tested, SWCNTs and MWCNTs were generally absent of mutagenic activity. Relatedly, there is a lack of reported mutagenic effects from testing the urine of laboratory subjects given large oral doses of SWCNTs and MWCNTs using the Ames assay. Aside from mutagenicity, some assays report cytotoxicity and some degree of DNA damage. In one of the few positive studies, an *in vitro* and *in vivo* study with MWCNTs significantly

---

[*] Tox21 is a collaboration between U.S.EPA, National Institutes of Environmental Health Sciences/National Toxicology Program, National Institutes of Health/National Human Genome Research Institute, NIH Chemical Genomics Center, and the Food and Drug Administration.

[†] The Ames assay is a method used to assess the mutagenic potential of test substances.

[‡] The Comet assay is a method for measuring DNA single- and double-strand breaks at the level of individual cells.

[§] The micronucleus assay is a test used in toxicological screening for potential genotoxic compounds.

[¶] The chromosome aberration assay is a method to determine chromosome breakage often induced by various types of mutagens.

induced micronuclei in A549 cells, enhanced the frequency of sister-chromatid exchange* in CHO AA8 cells, and induced DNA damage in the lungs of mice (Kato et al. 2012).

Assays for mutagenicity should be considered with caution, as these assays may not be suitable for detecting genotoxicity with nano-objects. For example, bacterial-based assays may not be able to introduce the material into the cell, which is an important requirement in some assays. The physicochemical properties of the tested materials are not reported, which leaves several considerations open for further study. For example, if the tested materials had been purified to remove metal contaminants, then these materials might provide results different than if the tested material was in its raw form.

The limited data retrieved thus far suggest that CNTs are not genotoxic or mutagenic. However, there is a question as to whether the assays for these end points are appropriate. Thus, this will likely be an area of continuing research. More research in these subject areas will help place the current results into a better context.

4.2.3.4 Carcinogenicity

In the short toxicological history of CNTs, one of the first observations was that CNTs resemble fibers. From a purely observational perspective, MWCNTs appear fiberlike when one describes their shape and size. They are often classified by their high aspect ratio. A number of studies have demonstrated that some CNTs can persist in the lungs. These are characteristics of materials that, for a toxicologist, would require well-designed experiments to address how these properties might be toxicologically significant. It was not much of a scientific step to compare the possible health effects of CNTs to those of asbestos fibers. It is uncertain if this is a fair or unfair comparison; however, the association has stuck.

The two studies that provide the most support for the fiberlike activity of CNTs are by Takagi et al. (2008) and Poland et al. (2008). Takagi and colleagues administered a high dose of either 3 mg MWCNTs/animal, crocidolite asbestos, or fullerenes via i.p. injection into the abdominal (peritoneal) cavity of a strain of mice (p53 +/− mice). Similarly, Poland and colleagues injected four types of MWCNTs via i.p. injection (50 µg dose) into the peritoneal cavity of mice. Takagi and colleagues reported that mice treated with MWCNTs (average width 100 nm, length between 100 and 20,000 nm, with 27.5% of particles longer than 5 µm) revealed moderate to severe fibrous peritoneal adhesions with slight ascites. Poland and

---

* Sister chromatid exchange is the exchange of genetic material between two identical sister chromatids.

colleagues reported inflammatory response and the formation of granulo-mas *in vivo* following i.p. administration, but they also promote the devel-opment of mesothelioma in more long-term *in vivo* studies.

As noted previously, these experiments need to be interpreted with caution for a number of reasons. First, the relevant route of exposure to humans is inhalation, not intraperitoneal injection. The exposure route for injection versus inhalation has different anatomies and physiologies, which will affect the absorption, transportation, metabolism, and excre-tion in the body. Second, the species of laboratory animal used by Takagi and colleagues (2008) is particularly sensitive to carcinogens because of the genetically modified inability of this mouse to repair genotoxic dam-age and/or to clear damaged cells through apoptosis.

Using intrascrotal exposures, Sakamoto and colleagues reported that MWCNTs possess carcinogenicity and induce mesothelioma in rats after exposure to a single injection (Sakamoto et al. 2009).

In summary, the data are very limited, but thus far suggest that CNTs are not carcinogenic via inhalation. Some CNTs demonstrate mesothelial responses that are, according to the authors, similar to fiberlike reactions when injected into the bodies of rats and mice. This will likely be an area of continued research, as some research reports DNA damage from some types of CNT exposures. It is likely that more research will help place the current results into better context.

4.2.4 Target Organ Toxicity

4.2.4.1 General Observations

Much of what is noted in this research is what possible hazards may exist due to exposure to CNTs. In other words, if one disregards dose and exposure from a toxicological assessment, then all a toxicologist will learn is the kind of adverse effects that can occur. But for a proper toxi-cological assessment, dose and exposure are fundamental components. Considerably more research is needed regarding dose and exposure, as many of the following experiments are not likely relevant to environmen-tal conditions. A large number of *in vivo* studies report on the acute and subchronic effects of CNTs and the formation of fibrosis—an abnormal formation of fibrous tissue—and granulomas (Taber and Thomas 1997). Lam and colleagues (2004) reported that agglomerated SWCNTs induced the formation of granulomas in instilled mice and, in some cases, intersti-tial inflammation in a dose-dependent manner, regardless of the metals present (nickel, iron, yttrium as catalyst) and their amount. In this experi-ment, purified (lack of metals) and unpurified (raw) materials were tested among other materials such as carbon black and quartz. Extrinsic factors may also play a role in the subject's response. Shvedova and colleagues

(2007) reported that SWCNT-induced changes were significantly more pronounced in the lungs of vitamin E–deficient animals than in mice fed a control diet.

The role of metal catalysts in nanomaterial toxicity is uncertain at this point. In some experiments, metal catalysts appear to contribute to toxicity, and in some experiments they do not. This will be an important continuing area of research. The structure of the molecule may also promote fibroblast cell growth. Wang and colleagues (2010) suggest that the dispersion status and size of the SWCNT structures is a critical factor in determining nanoparticle fibrogenicity and that matrix metalloproteinases may be involved in the fibrogenic process.

4.2.4.2 Acute Toxicity Testing

Acute toxicity tests are tests where doses are given in a single exposure, often high enough to produce a short onset of adverse effects. The traditional marker for acute toxicity testing is death. Comparing LD_{50} or LC_{50} across chemicals—given the same route often with the same test species—provides a level of understanding on the general short-term toxicity. This assumes that the experimental design is sufficient (e.g., dose adequate) and that the database of reliable studies is adequate to make dependable conclusions. There are few toxicity studies that have been retrieved from the literature. The general results of these studies are that, barring a significant concentration of metal catalysts, the acute oral toxicity for SWCNTs and MWCNTs is low. The following discussion focuses on results from the literature.

Warheit and colleagues (2004) reported that with a single pulmonary exposure via intratracheal instillation and pharyngeal aspiration to mice and rats, CNT exposure, in general, results in an acute, neutrophil-driven inflammatory and fibrotic response, with granuloma development associated with CNT aggregates. These effects were noted about two weeks after the one-time exposure. SWCNTs (lengths of 1 nm × 1–2 μm) were usually more potent than MWCNTs, at comparable doses, in inducing inflammation and granulomas. Warheit and colleagues suspected that catalytic metal impurities might play a role in toxicity. Mortality was noted following intratracheal instillation of high doses (16.7 mg/kg in mouse; 5 mg/kg in rat) of SWCNTs. The researchers believe that mortality is explained by congestion of the airways, leading to asphyxiation rather than toxicity from the failure of vital organ systems.

Regarding MWCNTs, Eillinger-Ziegelbauer and Pauluhn (2009) reported that doses as high as 241 mg/m³ MWCNTs for a single six-hour nose-only inhalation exposure of rats, followed by a three-month postexposure period, induced pulmonary inflammation. The researchers reported a dose-dependent effect and evidence that the response abated over time.

What appears to be noteworthy is that when the results of inflammation were compared and contrasted to a positive control—α-quartz at the same mass—MWCNTs induced a less pronounced inflammation. No mortality was observed at this dose and, thus, based on this experiment, the LC_{50} for these particular MWCNTs was greater than 241 mg/m^3.

For slightly longer periods of toxicity testing (e.g., exposures of 5 to 6 hours per day for 7 to 24 days of exposure), there are reports of testing both SWCNTs and MWCNTs in rodents. These testing periods have exposure periods that are weeks long but, in general, are of shorter duration than subchronic testing (e.g., 90 days). The term *subacute testing* is sometimes used to describe this sort of testing. The results of such tests are worthy of further discussion.

Li and colleagues (2007) reported that the lung responses to inhalation of SWCNTs include inflammation changes. Subjects exposed to inhalation of MWCNTs showed moderate proliferation and thickening of alveolar walls. In some cases, intratracheal instillation appeared to cause more severe effects than via inhalation exposure. The study by Li et al. reported that after two weeks of inhalation exposure, MWCNTs were engulfed by alveolar macrophages and the MWCNTs were distributed throughout the lung. No increases in inflammatory cell infiltration were found, with a lack of inflammation, granuloma formation, fibrosis, and tissue injury.

Mitchell and colleagues (2007) reported a rather interesting effect on systemic immunity by MWCNTs. This appeared for all doses of the test material. Spleen-derived cells showed suppressed T cell–dependent antibody response, decreased proliferation of T cells following mitogen stimulation, and altered natural killer (NK) cell killing. At least one experiment reported that immunological suppression continued for thirty days after exposure stopped.

The data suggest that CNTs are not acutely toxic unless substantial doses are given, often higher than would be likely from environmental exposures. Some of the tests in rodents discuss concentrations that were high enough to occlude respiratory structures in the laboratory rodent, causing death from asphyxiation rather than toxicity. Tests where testing material was given to the subject for longer periods of time demonstrate that inflammatory responses appear frequently for the doses provided. There might be a difference between inhalation and instillation exposures and pulmonary reactions. One important observation after several experiments is that after acute inhalation toxicity testing, some CNTs were noted to persist within the lungs for several months. CNTs have a relatively long half-life, which might be significant from an environmental health perspective. Thus, more testing needs to be done to determine which type of CNTs might induce greater risks.

4.2.4.3 Subchronic Inhalation Exposure

Subchronic toxicity inhalation studies are common. For example, U.S.EPA, as a part of the application for premanufacturing notification (PMN) under the Toxic Substances Control Act (TSCA), regularly performs them. The U.S.EPA has established testing guidelines for these ninety-day subchronic inhalation studies, and requires manufacturers of CNTs to perform them.

Two studies retrieved from the literature provide some information on the outcome of testing MWCNTs. For raw MWCNTs with metal oxides (10%), aerosol concentrations in the atmosphere ranging from 0.1 to 2.5 mg/m^3, and thirteen weeks of exposure, no pathological responses in major organs were noted. However, at doses greater than 0.1 mg/m^3, the lungs of rats had increased lung weights, granulomatous and neutrophilic inflammation, and intra-alveolar lipoproteinosis. At the lowest exposure level, 0.1 mg/m^3, rats had minimal granulomatous-type inflammation in the lungs and lung-associated lymph nodes, and these findings were considered by the authors to be subclinical and unlikely to be associated with functional effects. However, 0.1 mg/m^3 was considered a lowest observed adverse effect concentration (LOAEC) (Ma-Hock et al. 2009).

In a second subchronic inhalation study, Pauluhn (2010) exposed rats to MWCNTs via nose-only chambers for six hours per day and five days per week for thirteen consecutive weeks. The atmosphere the rats were exposed to ranged from 0.1 to 6 mg/m^3. Exposures above 0.4 mg/m^3, in a dose-related response, reported adverse effects. In the lungs of the subjects at these higher experimental exposures, MWCNTs caused elevations in certain types of white blood cells (observed from bronchoalveolar lavage) and microscopic lesions in the upper and the lower respiratory tracts. At the lowest tested concentration of 0.1 mg/m^3, all end points examined were unchanged from the control treatment group, and this concentration could be considered a no observed adverse effect concentration (NOAEC) (Pauluhn 2010).

The 0.1-mg/m^3 dose in the Ma-Hock study is a LOAEC, and it is a NOAEC in the Pauluhn study (Ma-Hock et al. 2009; Pauluhn 2010). This discrepancy might be explained by a number of different aspects, including the physicochemical properties of MWCNTs, the experimental exposure scenarios, or that the LOAEC for the one study, while effects were observed, was not clinically relevant. Further studies of MWCNTs are likely to provide more information as the research community continues to publish its findings.

4.2.4.4 Other Organs

Regarding studies on other organ systems, results were retrieved on skin and some reproductive end points. While not a toxicological study, one study reports potential occupational dermal exposure to CNTs. SWCNTs were measured on gloves that were worn while performing a work task (Maynard et al. 2004). Estimates of the SWCNT deposit during production and handling ranged from 217 to 6020 µg (0.2–6 mg) per hand, with most of the material appearing on the parts of the gloves in direct contact with surfaces (i.e., inner surfaces of fingers and palms). There is one short-term dermal study in the literature. Mice were dermally exposed to raw SWCNTs (1.5 nm × 1–100 µm; 30 wt% Fe) daily for five days at doses ranging from 40 to 160 µg/mouse. At the highest dose, a significant increase in skin bifold thickness (150%) was observed. Skin bifold thickness is one possible measure of the development of inflammation. Histologic examination of the skin revealed that an increase in fibroblasts occurred at the two highest doses (Murray et al. 2009).

There are a few studies regarding dermal irritation or sensitization and MWCNTs. Two different sizes of MWCNTs with less than 0.1% metal content were tested for their dermal and eye irritation potential *in vivo* and *in vitro*. A dose of 0.5 mg MWCNTs applied for four hours on a clipped skin area of rabbits did not induce any dermal reactions such as erythema and edema until seventy-two hours postexposure (Kishore et al. 2009).

Regarding reproductive studies, Bai and colleagues (2010) administered single or repeated i.v. injections (5 mg/kg) of water-soluble functionalized MWCNTs for up to ninety days in male mice. The MWCNTs might be useful agents for imaging applications. Bai and colleagues reported that fertility was unaffected (no significant effect on sperm, selected hormones, pregnancy rate, and delivery success of the mated females).

Cheng, Flahaut, and Cheng (2007) reported on the reproductive effects of raw SWCNTs and MWCNTs in a nonmammalian species, the zebra fish. They noted the presence of metal catalysts in some of the testing materials. There were lower survival rates in the second generation and a significant hatching delay in zebra fish embryos. It is not certain whether this effect was due to the CNTs or the metal impurities. A difference between SWCNTs and MWCNTs was that embryonic chorions prevented SWCNTs from exposing the embryo and its associated structures.

Very few studies in the literature have been devoted to investigation of dermal and reproductive end points. For dermal end points, CNTs appear to be rather benign, aside from irritation, for the doses administered and the experimental conditions. There is not enough literature related to CNTs and reproductive toxicity to make any overall conclusions at this time. There is interest in using different animal models, and the zebra fish continues to gain recognition as a viable model. Future research is expected to examine the issue of reproductive end points.

4.3 Future Directions

Nanotechnology appears to offer incalculable opportunities for the future. As noted in this chapter, with the focus on CNTs, there are possible hazards as well as the possibility for some nano-objects to enter the environment. A product can enter the environment at many stages of its life cycle, mainly through contact with consumers and industrial workers, but also through the disposal of waste materials. Scientists are working to address these issues. The prevailing spirit in the science world is to learn from the industrial chemical releases and accidents of the past—to "get it right" this time. The International Standards Organization has developed documents that allow manufacturers to handle nano-objects safely and has developed a list of physicochemical characteristics that are deemed important to measure when conducting toxicological tests. Some corporations have developed internal guidelines for safety. Industry has taken an aggressive approach by developing worker safety practices that include personal protective equipment and ventilation controls in the workplace. In the United States, the National Institute for Occupational Safety and Health has a dedicated field team of nanotechnology safety and health experts that visits companies on a voluntary basis, reviews their internal practices, and conducts occupational-related sampling (NIOSH 2008). New tools are being developed by companies and research institutions to aid in the creation of nano-related toxicological information (ZNEP 2010; RTI 2012). The concept of Green Chemistry—the design of chemical products and processes that reduce or eliminate the use or generation of hazardous substances—is prominent in the nanotechnology arena as well (U.S.EPA 2012). Research centers, such as the Safer Nanomaterials and Nanomanufacturing Initiative (SNNI) in Oregon, are working to develop safer alternative materials (SNNI 2012).

The experimental work in the field of nanotechnology is a mixed bag. Some of the research has been conducted satisfactorily, but there are a number of studies in the peer-reviewed literature that cannot be used in the assessment of environmental health and safety. In general, the concerns include the lack of strong experimental designs and the lack of physicochemical characterization of the administered materials. Relevant dose–response data are critical research needs. At this point, we have summarized the best science from the group reviewed; however, that does not mean that the studies are uniformly equivalent in reliability of their data for risk assessment purposes. The literature contains far more substantial articles than when the first edition of this book was published in 2008, but more quality work is needed.

Hopefully, good scientific studies will continue to be conducted. The money needed to conduct research is scarce, and thus all scientists should strive to develop well-designed and relevant experiments, perform proper characterization of nano-objects, and report all information, including negative data, so that the toxicological community knows where to focus its research.

Acknowledgments

I thank Mr. Elmer Diaz, who both collected and helped me analyze the extensive toxicological literature on carbon nanotubes. I thank Mr. Gavin Bell for his assistance in preparing the various versions of this manuscript. I thank Ms. Kerry King for her assistance in editing and library work.

References

Aiso, S., K. Yamazaki, Y. Umeda, M. Asakura, T. Kasai, M. Takaya, T. Toya, et al. 2010. Pulmonary toxicity of intratracheally instilled multiwall carbon nanotubes in male Fischer 344 rats. *Ind. Health* 48(6): 783–95.

Asakura, M., T. Sasaki, T. Sugiyama, M. Takaya, S. Koda, K. Nagano, H. Arito, and S. Fukushima. 2010. Genotoxicity and cytotoxicity of multi-wall carbon nanotubes in cultured Chinese hamster lung cells in comparison with chrysotile A fibers. *J. Occup. Health* 52(3): 155–66.

Bai, Y., Y. Zhang, J. Zhang, Q. Mu, W. Zhang, E. R. Butch, S. E. Snyder, and B. Yan. 2010. Repeated administrations of carbon nanotubes in male mice cause reversible testis damage without affecting fertility. *Nat. Nanotechnol.* 5(9): 683–89.

Bottini, M., S. Bruckner, K. Nika, N. Bottini, S. Bellucci, A. Magrini, A. Bergamaschi, and T. Mustelin. 2006. Multi-walled carbon nanotubes induce T lymphocyte apoptosis. *Toxicol. Lett.* 160(2): 121–26.

Cheng, J., E. Flahaut, and S. H. Cheng. 2007. Effect of carbon nanotubes on developing zebrafish (*Danio rerio*) embryos. *Environ. Toxicol. Chem.* 26(4): 708–16.

Cherukuri, P., C. J. Gannon, T. K. Leeuw, H. K. Schmidt, R. E. Smalley, S. A. Curley, and R. B. Weisman. 2006. Mammalian pharmacokinetics of carbon nanotubes using intrinsic near-infrared fluorescence. *Proc. Natl. Acad. Sci. USA* 103(50): 18882–86.

Chou, C. C., H. Y. Hsiao, Q.-S. Hong, C.-H. Chen, Y.-W. Peng, H.-W. Chen, and P.-C. Yang. 2008. Single-walled carbon nanotubes can induce pulmonary injury in mouse model. *Nano Lett.* 8(2): 437–45.

Donaldson, K., V. Stone, A. Clouter, L. Renwick, and W. MacNee. 2001. Ultrafine particles. *Occup. Environ. Med.* 58(3): 211–16.

Dumortier, H., S. Lacotte, G. Pastorin, R. Marega, W. Wu, D. Bonifazi, J.-P. Briand, M. Prato, S. Muller, and A. Bianco. 2006. Functionalized carbon nanotubes are non-cytotoxic and preserve the functionality of primary immune cells. *Nano Lett.* 6(7): 1522–28.

Ellinger-Ziegelbauer, H., and J. Pauluhn. 2009. Pulmonary toxicity of multi-walled carbon nanotubes (Baytubes) relative to alpha-quartz following a single 6h inhalation exposure of rats and a 3 months post-exposure period. *Toxicology* 266(1–3): 16–29.

Environmental Defense and DuPont Nano Partnership. 2007. *Nano risk framework.* Dover, DE: DuPont.

Folkmann, J. K., L. Risom, N.R. Jacobsen, H. Wallin, S. Loft, and P. Møller. 2009. Oxidatively damaged DNA in rats exposed by oral gavage to C60 fullerenes and single-walled carbon nanotubes. *Environ. Health Perspect.*

Howard, J., C. L. Geraci, and P. Schulte. 2009. Occupational disease and nanoparticles. *NIOSH Science Blog.* Atlanta, GA: Centers for Disease Control. http://blogs.cdc.gov/niosh-science-blog/2009/08/nano-4/

ISO. 2008. Nanotechnologies: Terminology and definitions for nano-objects—nanoparticle, nanofibre and nanoplate. ISO/TS 27687. Geneva, Switzerland: International Organization for Standards.

———. 2010. Nanotechnologies: Vocabulary. ISO/TS 80004-3:2010. Geneva, Switzerland: International Organization for Standards.

———. 2012. Nanotechnologies: Guidance on physico-chemical characterization of engineered nanoscale materials for toxicologic assessment. ISO/DTR 13014. Geneva Switzerland: International Organization for Standards.

Jacobsen, N., R. G. Pojana, P. White, P. Møller, C. A. Cohn, K. S. Korsholm, U. Vogel, A. Marcomini, S. Loft, and H. Wallin. 2008. Genotoxicity, cytotoxicity, and reactive oxygen species induced by single-walled carbon nanotubes and C(60) fullerenes in the FE1-Mutatrade markMouse lung epithelial cells. *Environ. Mol. Mutagen* 49(6): 476–87.

Jia, G., H. Wang, L. Yan, X. Wang, R. Pei, T. Yan, Y. Zhao, and X. Guo. 2005. Cytotoxicity of carbon nanomaterials: Single-wall nanotube, multi-wall nanotube, and fullerene. *Environ. Sci. Technol.* 39(5): 1378–83.

Johnston, D. E., M. F. Islam, A. G. Yodh, and A. T. Johnson. 2005. Electronic devices based on purified carbon nanotubes grown by high-pressure decomposition of carbon monoxide. *Nat. Mater.* 4(8): 589–92.

Kagan, V. E., N. V. Konduru, W. Feng, B. L. Allen, J. Conroy, Y. Volkov, I. I. Vlasova, et al. 2010. Carbon nanotubes degraded by neutrophil myeloperoxidase induce less pulmonary inflammation. *Nat. Nanotechnol.* 5(5): 354–59.

Kagan, V. E., Y. Y. Tyurina, V. A. Tyurin, N. V. Konduru, A. I. Potapovich, A. N. Osipov, E. R. Kisin, et al. 2006. Direct and indirect effects of single walled carbon nanotubes on RAW 264.7 macrophages: Role of iron. *Toxicol. Lett.* 165(1): 88–100.

Kapralov, A. A., N. Yanamala, W. H. Feng, B. Fadeel, A. Star, A. A. Shvedova, and V. E. Kagan. 2011. Biodegradation of carbon nanotubes by eosinophil peroxidase. Society of Toxicology Annual Meeting 2011, Washington, DC.

Kato, T., Y. Totsuka, K. Ishino, Y. Matsumoto, Y. Tada, D. Nakae, S. Goto, et al. 2012. Genotoxicity of multi-walled carbon nanotubes in both in vitro and in vivo assay systems. *Nanotoxicology*. http://www.ncbi.nlm.nih.gov/pubmed/22397533

Kishore, A., S. P. Surekha, and P. B. Murthy. 2009. Assessment of the dermal and ocular irritation potential of multi-walled carbon nanotubes by using in vitro and in vivo methods. *Toxicol. Lett.* 191(2–3): 268–74.

Kisin, E. R., A. R. Murray, M. J. Keane, X.-C. Shi, D. Schwegler-Berry, O. Gorelik, S. Arepalli, et al. 2007. Single-walled carbon nanotubes: Geno- and cytotoxic effects in lung fibroblast V79 cells. *J. Toxicol. Environ. Health A* 70(24): 2071–79.

Kisin, E. R., A. R. Murray, L. Sargent, D. Lowry, M. Chirila, K. J. Siegrist, D. Schwegler-Berry, et al. 2011. Genotoxicity of carbon nanofibers: Are they potentially more or less dangerous than carbon nanotubes or asbestos? *Toxicol. Appl. Pharmacol.* 252(1): 1–10.

Köhler, A. R., C. Som, A. Helland, and F. Gottschalk. 2007. Studying the potential release of carbon nanotubes throughout the application life cycle. *Journal of Cleaner Production* 16(8–9): 927–37.

Lam, C. W., J. T. James, R. McCluskey, and R. L. Hunter. 2004. Pulmonary toxicity of single-wall carbon nanotubes in mice 7 and 90 days after intratracheal instillation. *Toxicol. Sci.* 77(1): 126–34.

Li, J. G., W. X. Li, J. Y. Xu, X. Q. Cai, R. L. Liu, Y. J. Li, Q. F. Zhao, and Q. N. Li. 2007. Comparative study of pathological lesions induced by multiwalled carbon nanotubes in lungs of mice by intratracheal instillation and inhalation. *Environ. Toxicol.* 22(4): 415–21.

Ma-Hock, L., S. Treumann, V. Strauss, S. Brill, F. Luizi, M. Mertler, K. Wiench, A. O. Gamer, B. van Ravenzwaay, and R. Landsiedel. 2009. Inhalation toxicity of multiwall carbon nanotubes in rats exposed for 3 months. *Toxicol. Sci.* 112(2): 468–81.

Manna, S. K., S. Sarkar, J. Barr, K. Wise, E. V. Barrera, O. Jejelowo, A. C. Rice-Ficht, and G. T. Ramesh. 2005. Single-walled carbon nanotube induces oxidative stress and activates nuclear transcription factor-κB in human keratinocytes. *Nano Lett.* 5(9): 1676–84.

Maynard, A. D., P. A. Baron, M. Foley, A. A. Shvedova, E. R. Kisin, and V. Castranova. 2004. Exposure to carbon nanotube material: Aerosol release during the handling of unrefined single-walled carbon nanotube material. *J. Toxicol. Environ. Health A* 67(1): 87–107.

McNeil, S. E. 2009. Nanoparticle therapeutics: A personal perspective. *Wiley Interdisp. Rev. Nanomed. Nanobiotechnol.* 1(3): 264–71.

Mitchell, L. A., J. Gao, R. Vander Wal, A. Gigliotti, S. W. Burchiel, and J. D. McDonald. 2007. Pulmonary and systemic immune response to inhaled multiwalled carbon nanotubes. *Toxicol. Sci.* 100(1): 203–14.

Muller, J., F. Huaux, N. Moreau, P. Misson, J.-F. Heilier, M. Delos, M. Arras, A. Fonseca, J. B. Nagy, and D. Lison. 2005. Respiratory toxicity of multi-wall carbon nanotubes. *Toxicol. Appl. Pharmacol.* 207(3): 221–31.

Murray, A. R., E. Kisin, S. S. Leonard, S. H. Young, C. Kommineni, V. E. Kagan, V. Castranova, and A. A. Shvedova. 2009. Oxidative stress and inflammatory response in dermal toxicity of single-walled carbon nanotubes. *Toxicology* 257(3): 161–71.

Needle, D. 2006. Nano spray recall raises potential health risks. *Internet News.com*, April 14.

NIOSH. 2008. *NIOSH nanotechnology field research effort*. Atlanta, GA: Centers for Disease Control and Prevention.

———. 2010. *Occupational exposure to carbon nanotubes and nanofibers*. Atlanta, GA: Department of Health and Human Services.

NRC, 1983. *Risk assessment in the federal government: Managing the process*. Washington, DC: National Academy Press for National Research Council.

Oberdörster, G., A. Maynard, K. Donaldson, V. Castranova, J. Fitzpatrick, K. Ausman, J. Carter, et al. 2005. Principles for characterizing the potential human health effects from exposure to nanomaterials: Elements of a screening strategy. *Part Fibre Toxicol.* 2:8.

Oberdörster, G., E. Oberdörster, and J. Oberdörster. 2005. Nanotoxicology: An emerging discipline evolving from studies of ultrafine particles. *Environ. Health Perspect.* 113(7): 823–39.

OECD. 2010. *Compilation of nanomaterial exposure mitigation guidelines relating to laboratories*. Geneva, Switzerland: Organization for Economic Cooperation and Development.

Pacurari, M., X. J. Yin, J. Zhao, M. Ding, S. S. Leonard, D. Schwegler-Berry, B. S. Ducatman, et al. 2008. Raw single-wall carbon nanotubes induce oxidative stress and activate MAPKs, AP-1, NF-kappaB, and Akt in normal and malignant human mesothelial cells. *Environ. Health Perspect.* 116(9): 1211–7.

Patlolla, A. K., S. M. Hussain, J. J. Schlager, S. Patlolla, and P. B. Tchounwou. 2010. Comparative study of the clastogenicity of functionalized and nonfunctionalized multiwalled carbon nanotubes in bone marrow cells of Swiss-Webster mice. *Environ. Toxicol.* 25(6): 608–21.

Pauluhn, J. 2010. Multi-walled carbon nanotubes (Baytubes): Approach for derivation of occupational exposure limit. *Regul. Toxicol. Pharmacol.* 57(1): 78–89.

Poland, C. A., R. Duffin, I. Kinloch, A. Maynard, W. A. H. Wallace, A. Seaton, V. Stone, S. Brown, W. MacNee, and K. Donaldson. 2008. Carbon nanotubes introduced into the abdominal cavity of mice show asbestos-like pathogenicity in a pilot study. *Nat. Nanotechnol.* 3:423–28.

Porter, D. W., A. F. Hubbs, R. R. Mercer, N. Wu, M. G. Wolfarth, K. Sriram, S. Leonard, et al. 2010. Mouse pulmonary dose- and time course-responses induced by exposure to multi-walled carbon nanotubes. *Toxicology* 269(2–3): 136–47.

Rogers, B., S. Pennathur, and J. Adams. 2011. *Nanotechnology: Understanding Small Systems*. London: CRC Press.

Royal Society. 2004. *Nanoscience and nanotechnologies: Opportunities and Uncertainties*. Cardiff, UK: Clyvedon Press.

RTI. 2012. *Environmental and biological concepts of the nanomaterials registry*. Research Triangle Park, NC: RTI International.

Ryman-Rasmussen, J. P., M. F. Cesta, A. R. Brody, J. K. Shipley-Phillips, J. I. Everitt, E. W. Tewksbury, O. R. Moss, et al. 2009. Inhaled carbon nanotubes reach the subpleural tissue in mice. *Nat. Nanotechnol.* 4:747–51.

Sakamoto, Y., D. Nakae, N. Fukumori, K. Tayama, A. Maekawa, K. Imai, A. Hirose, T. Nishimura, N. Ohashi, and A. Ogata. 2009. Induction of mesothelioma by a single intrascrotal administration of multi-wall carbon nanotube in intact male Fischer 344 rats. *J. Toxicol. Sci.* 34(1): 65–76.

Sharma, C. S., S. Sarkar, A. Periyakaruppan, J. Barr, K. Wise, R. Thomas, B. L. Wilson, and G. T. Ramesn. 2007. Single-walled carbon nanotubes induces oxidative stress in rat lung epithelial cells. *J. Nanosci. Nanotechnol.* 7(7): 2466–72.

Shvedova, A. A., V. Castranova, E. Kisin, D. Schwegler-Berry, A. Murray, V. Gandelsman, A. Maynard, and P. Baron. 2003. Exposure to carbon nanotube material: assessment of nanotube cytotoxicity using human keratinocyte cells. *J. Toxicol. Environ. Health A* 66(20): 1909–26.

Shvedova, A. A., E. R. Kisin, A. R. Murray, O. Gorelik, S. Arepalli, V. Castranova, S. H. Young, et al. 2007. Vitamin E deficiency enhances pulmonary inflammatory response and oxidative stress induced by single-walled carbon nanotubes in C57BL/6 mice. *Toxicol. Appl. Pharmacol.* 221(3): 339–48.

Singh, R., D. Pantarotto, L. Lacerda, G. Pastorin, C. Klumpp, M. Prato, A. Bianco, and K. Kostarelos. 2006. Tissue biodistribution and blood clearance rates of intravenously administered carbon nanotube radiotracers. *Proc. Natl. Acad. Sci. USA* 103(9): 3357–62.

SNNI. 2012. Safer nanomaterials and nanomanufacturing initiative. Eugene, OR: Materials Science Institute, University of Oregon. http://www.greennano.org/

Song, Y., X. Li, and X. Du. 2009. Exposure to nanoparticles is related to pleural effusion, pulmonary fibrosis and granuloma. *Eur. Respir. J.* 34(3): 559–67.

Taber, C. W., and C. L. Thomas. 1997. *Taber's Cyclopedic Medical Dictionary*. Philadelphia: F.A. Davis.

Takagi, A., A. Hirose, T. Nishimura, N. Fukumori, A. Ogata, N. Ohashi, S. Kitajima, and J. Kanno. 2008. Induction of mesothelioma in p53+/− mouse by intraperitoneal application of multi-wall carbon nanotube. *J. Toxicol. Sci.* 33(1): 105–16.

U.S.EPA. 1976. Toxic Substances Control Act of 1976. Washington, DC: U.S. Environmental Protection Agency.

———. 1997. *Chemistry assistance manual for premanufacture notification submitters*. Washington, DC: Office of Pollution, Prevention and Toxicology, U.S. Environmental Protection Agency.

———. 1998. *Health effects test guidelines*. OPPTS 870.1300. Washington, DC: U.S. Environmental Protection Agency.

———. 2011. *Ultrafine particle research*. Washington, DC: U.S. Environmental Protection Agency.

———. 2012. *Green chemistry*. Washington, DC: U.S. Environmental Protection Agency.

Wang, L., R. R. Mercer, Y. Rojanasakul, A. Qiu, Y. Lu, J. F. Scabilloni, N. Wu, and V. Castranova. 2010. Direct fibrogenic effects of dispersed single-walled carbon nanotubes on human lung fibroblasts. *J. Toxicol. Environ. Health A* 73(5): 410–22.

Warheit, D. B. 2010. Debunking some misconceptions about nanotoxicology. *Nano Lett.* 10(12): 4777–82.

Warheit, D. B., B. R. Laurence, K. L. Reed, D. H. Roach, G. A. M. Reynolds, and T. R. Webb. 2004. Comparative pulmonary toxicity assessment of single-wall carbon nanotubes in rats. *Toxicol. Sci.* 77(1): 117–125.

Wick, P., P. Manser, L. K. Limbach, U. Dettlaff-Weglikowska, F. Krumeich, S. Roth, W. J. Stark, and A. Bruinink. 2007. The degree and kind of agglomeration affect carbon nanotube cytotoxicity. *Toxicol. Lett.* 168(2): 121–31.

Yu, M.-F., O. Lourie, M. J. Dyer, K. Moloni, T. F. Kelly, and R. S. Ruoff. 2000. Strength and breaking mechanism of multiwalled carbon nanotubes under tensile load. *Science Magazine* 287(5453): 637–40.

ZNEP. 2010. Zurich introduces new risk assessment tool for nanotechnology. *Online Techfocus Articles,* Zurich Technology Insurance Services, Zurich, Switzerland. http://www.techinsurance.zurichna.com/?p=1,2,1,1&prid=145

Further Reading

Fiorino, D. 2010. *Voluntary Initiatives, Regulation, and Nanotechnology Oversight: Charting a Path.* Washington, DC: Woodrow Wilson International Center for Scholars, The Pew Charitable Trusts.

NanoEHS: The Virtual Journal of Nanotechnology Environment, Health and Safety, all issues; found online. International Council on Nanotechnology, Rice University, Houston, TX.

OSHA. 2008. *Health Effects and Workplace Assessments and Controls.* Website of Occupational Safety and Health Administration, U.S. Department of Labor, Washington, DC.

Stone, V. 2009. *Engineered nanoparticles: Review of health and environmental safety final report.* Seventh Framework Programme of the European Union, Brussels, Belgium.

5

The State of the Science: Environmental Risks

Jo Anne Shatkin

CONTENTS

Applying the general prefix nano does not in itself automatically render a material harmful.

Bernd Nowack

Even in the built environment, reliance on water, air, and natural resources is the basis of the high quality of life people aim to enjoy in modern society. Everyone requires clean water for drinking, cooking, bathing, irrigation, and recreation. When the air is polluted, it affects everyone's breathing, in some people stimulating respiratory distress and affecting heart functioning. The food supply is impacted when land and water resources are contaminated. As humans, we exist as part of Earth's ecosystem—as animals in the food chain. Humans are not quite at the top, but we are near it, particularly urban dwellers. We rely on all of the lower levels of the ecosystem to function in order for us to survive.

There are many reasons to be concerned about whether nanomaterials affect the environment, not the least of which is stewardship. Scientists and regulators study the effects of substances on the environment to assess impacts on specific populations or representative organisms as indicators of the extent of effects. This information is used to make decisions about regulating substances to protect public health and the environment. That is, to protect human health as well as the air, land, and water in which we exist, we must ensure that our activities are responsible, that they maintain or improve our existence, and that they preserve health for future generations.

Beyond the food chain, people are impacted by the health of plants and animals in many ways. Consider the following issues:

1. Environmental contamination can change the health of the world around us, and thus the resources humans depend on.

2. People rely on the environment, and because human activities often affect the environment, attention to the impacts of development on the world around us is necessary.

3. Ecological impacts can be sentinels—that is, observation of toxicity to fish and other aquatic organisms can be an indicator of adverse effects on people.

4. With global trade, the world's food supply is complex and relies on technology and food grown in different parts of the world that can be impacted if pollution or disease outbreaks affect the complex food web of an ecosystem, with unpredictable effects.

5. There is a lack of equity between the wealthy who enjoy a clean environment and those who are poor and live in a dirty environment. Protecting the environment is also a social justice issue.

It is difficult to characterize the state of knowledge about the environmental effects of nanomaterials. There is a richer database than in 2007, but significant gaps remain, and the scientific understanding of the behavior of specific nanomaterials in the environment is dynamic and diffuse. Defining the state of the science in twenty pages or so inherently means that this discussion is a 30,000-foot flyover. As discussed previously, nanomaterials are diverse, and it is not currently possible to extrapolate the findings from one nanomaterial to another. So, describing the environmental risks of "nanomaterials" is an exercise akin to describing the attributes of "food." Necessarily, this chapter will focus on identifying some of the key concerns and recent findings about the behaviors of nanomaterials entering the environment, and present a few well-studied examples as illustration.

Unlike people, the diversity of environmental "receptors" is much broader, existing in different habitats, either on land, in fresh or salt water, estuaries, etc. Some species are more sensitive than others, and they live in ecosystems within a diversity of species at differing levels of organization. In soils, for example, there are bacteria and other microbes, plants, insects and other invertebrates, as well as birds and mammals. Environmental studies of the effects of substances will often look at several of these types of organisms, or in other cases will only test specific species as indicators.

Many organizations have developed standardized tests to evaluate the environmental toxicity of chemical substances. These tests evaluate the effects of chemicals at different concentrations to assess how toxic they may be to different species.

The Organization for Economic Cooperation and Development (OECD) has developed thirty-five standardized methods for testing the effects of substances on the biological environment, including short-term, or acute, toxicity tests and longer term reproductive, growth, and survival tests for organisms like earthworms, fish, soil microbes, birds, bees, and others. While these tests have not been validated for nanomaterials, there are several published studies using these assays to assess them.

The U.S. Environmental Protection Agency (U.S.EPA) publishes a variety of methods for substances it regulates and is increasingly working toward harmonization of these standards with others. For example, U.S.EPA has proposed using a standard method for estimating removal of chemicals during wastewater treatment for engineered nanomaterials (ENM), the Activated Sludge Sorption Isotherm test guideline OPPTS 835.1110 (U.S.EPA 2010a). As discussed later in this chapter, a recent study suggests that this test is not valid and that it underpredicts removal of ENM in wastewater treatment plants (Kiser et al. 2012). Several new assays have been developed and are being used to assess the biological effects of nanomaterials, for example, the embryonic zebra fish assay (Zhu et al. 2007; Usenko, Harper, and Tanguay 2007).

A number of published studies rely on *in vitro* biological assays that assess specific behavior in cells or chemical tests, outside of whole organisms. These types of studies are desirable because they are fast, inexpensive, and avoid animal testing. However, they do not account for the dynamic physiology of whole organisms and, to date, have not been well correlated with animal studies for nanomaterials. Later, a couple of examples of false results from *in vitro* studies are discussed.

The first major finding is that it is still not possible to extrapolate studies of one ENM to others. In fact, the same materials made by different preparations and manufacturers demonstrate different behavior in biological and environmental systems, such that each material must be individually studied to measure its effects. This is of course impractical, and there are not enough resources, in the form of funding, researchers, test animals, or regulatory reviewers, to effectively test every possible ENM formulation. Even if there were, the potential for harm does not rise to the level where this would make sense. There are many more prevalent health and environmental issues that deserve priority, such as climate change, cancer research, and preventable diseases that, if addressed, will save many more lives. Fortunately, there are ways to categorize the uses of ENM and prioritize what information is necessary to ensure public health and environmental protection while using them.

A second major finding is that the surface properties of ENMs underlie their environmental behavior. One key characteristic of ENMs is their relatively large surface area in comparison to the mass of the material. Because the particles are so small, proportionately more of each particle is on the surface. Many ENMs have very reactive surfaces, which attract other molecules to them. This means that it is rare to find individual ENM particles in the environment. More likely is that ENMs form aggregates or agglomerates of ten or more particles, or that they attract biological molecules, the so-called corona effect, or bind other types of particles and act as carriers (Scown, van Aerle, and Tyler 2010), sometimes called the Trojan horse phenomenon.

Further, it remains a pressing challenge to detect ENMs in environmental media. If one is interested in measuring engineered iron particles in soil or groundwater, how can one measure them against the very high background levels of iron? New methods have emerged in the past several years, and there are efforts at standardization, but it remains a challenge to understand the transport, fate, and toxicity of ENMs in the environment, in part because of the difficulty in adequately measuring nanoparticles in complex environments.

5.1 Antimicrobial Properties

Many types of nanoparticles have been investigated for antimicrobial properties. What is an antimicrobial property? It is the ability of a substance to kill or inhibit the growth of microbes or microbiological organisms. Microbes include bacteria, such as *Escherichia coli* and Salmonella, which cause gastrointestinal illness, and also *Lactobacillus acidophilus*, the active ingredient in yogurt that improves digestion. Bacteria tend to be microns in size (although nanobacteria are an emerging area of research), whereas viruses are also microbes, but are of nanoscale size.

Not all microbes cause diseases. Those that do are called pathogens. Pathogens can cause disease by infecting the food, water, and air that people eat, drink, and breathe. Microbes that are not disease causing (nonpathogenic organisms) are important and often necessary components in the digestive tract, in soil, and in complex ecosystems. Nonselective killing of microbes can be harmful to health and the environment. While it is good for public health to kill pathogens, killing non-disease-causing environmental microbes can disrupt the natural processes of complex food webs in which microbes play a vital role, for example, creating food for higher organisms and breaking down dead organic matter into soil components.

The development of nanoscale materials for use as antimicrobials applies the greater reactivity and surface area of the smaller particles for more effective killing of pathogens. Applications include fabric coatings, food packaging, kitchen products, water taps, food contact surfaces, medical instruments, and other consumer products such as sporting goods, electronics, and door knobs (WWCS 2011). There are benefits to antimicrobial products: Pathogens are not transferred from the treated surfaces, thereby increasing food safety and decreasing the likelihood of contracting an infection during hospital visits or from doors, handrails, waiting rooms, bus seats, and other surfaces in public places. If you are a poor cleaner, antimicrobial treatment of appliances and kitchen counters in your home can reduce the presence of pathogens and resulting illnesses.

Because they only target microbes, some nanoparticle antimicrobials may be less toxic than the current alternatives, such as the chemical triclosan, which is now in hundreds of products and may be hormonally active (Jacobs, Nolan, and Hood 2005; Veldhoen et al. 2006), disrupting the endocrine systems of larger animals, including people. Since not all microbes are pathogens, wide use of antimicrobial coatings on consumer products could have some unintended effects, for example the development of antimicrobial resistance to microbes, which has occurred as a result of the wide introduction of antimicrobials in soaps and cleaning products (Pruden et al. 2006). Some pathogens that cause common hospital infections are resistant to antibiotics as a result of overuse for medical purposes and in products, and these subsequently have a wide occurrence in the environment. Increasing the frequency with which antimicrobials are used will increase microbial resistance to antibiotics, decreasing our ability to treat infections and affect our immunity toward them. As a result, society is increasingly at risk of an outbreak of disease caused by antibiotic-resistant microbes. Generally, our immune systems can fight off these microbes, but there are sensitive subpopulations of immunodeficient people who cannot. These include people whose immune systems are weakened because of other illnesses and also those with immune system diseases including lupus, AIDS, and others. Increasingly, as antimicrobial resistance grows, so does the risk even to healthy people.

Studies evaluating the effects of nanoscale silver, fullerenes, titanium dioxide, and carbon nanotubes have identified antimicrobial properties associated with exposure. These types of nanoscale particles are now being tested and applied for use in numerous applications. By far, nanoscale silver is currently the most widely developed for antimicrobial use.

5.1.1 Antimicrobial Properties of Nanoscale Silver

Colloidal silver and silver ions have long been used as antimicrobial agents. Nanoscale silver particles have been increasingly used as an alternative to conventional silver, and various sizes and shapes are rapidly entering the market in a range of consumer products. According to the Nanosilver Working Group and a recent analysis (Nowack, Krug, and Height 2011), colloidal silver products contain nanoscale silver particles and have been used commercially since the late 1890s; some 320 tons per year are currently produced. Further, according to Nowack et al., the first silver pesticide registered under FIFRA in 1953 was nanosilver, and over 50% of U.S.EPA-registered silver antimicrobials may be nanosilver particles, although only 7% claim to use nanoparticles (Nowack et al. 2011).

Nanosilver is as effective as silver ions, which are soluble and not nanoparticles, in killing *E. coli* (Sondi and Salopek-Sondi 2004); efficacy varies with the shape of the particle (Pal, Tak, and Song 2007). The nanosilver particles interfere with and destabilize the outside membrane of a microbe such that the contents of the cell are released, which kills it. This is specifically a microbial effect; the nanosilver does not affect mammalian cell membranes, which is part of why nanosilver can be found in over 300 consumer products on the market today, compared with fewer than 100 in 2007 (WWCS 2011). Some commercial vendors claim that silver is not toxic to people, although the U.S. Food and Drug Administration has banned the use of colloidal silver for gastrointestinal treatment (FDA 1999) and has nominated nanosilver to be tested for mammalian toxicity by the National Toxicology Program (NTP 2007).

The Federal Insecticide, Fungicide, and Rodenticide Act (FIFRA) has some of the most extensive premarket testing requirements within U.S.EPA. In 2011, the Office of Pesticide Programs (OPP) released a conditional registration, regulating the first submission of a nanosilver product that was labeled as such. The decision was a conditional approval for the use of a nanoscale-silver fabric treatment. OPP also issued a draft Federal Register Notice to require that any producer of "intentionally produced" nanoscale materials report, under FIFRA Section 6(a)(2), their use in products, regardless of registration status. OPP defines nanoscale materials as: "an ingredient that contains particles that have been intentionally produced to have at least one dimension that measures between approximately 1 and 100 nanometers."

This is important because even if colloidal silver contains nanosilver particles, novel forms of nanosilver are being produced that contain unique properties as a function of their size and shape. For example, triangular nanosilver particles have been shown to have a greater toxicity and be more persistent than spherical particles (U.S.EPA 2010b).

Aquatic and lower level organisms are more sensitive to silver than mammals and humans. The toxicity of silver ions to bacteria and small aquatic organisms and fish is well known and is likely to be similar for nanosilver. However, recent studies suggest that nanoparticulate silver toxicity can be differentiated from the effects of silver ions (e.g., Asharani et al. 2008; Navarro et al. 2008). Silver nanoparticles are small enough to penetrate the cell membranes of zebra fish embryos. Once inside the cell, nanoparticles can disrupt cellular processes (Lee, Lim, and Kim 2007). Thus, it is important to understand the effects of nanoparticles on environmental organisms and ecosystems. While water chemistry may convert nanoparticles of silver to other forms, the novel types of nanosilver particles may behave differently than those produced commercially over the last century.

Laboratory studies measure the impacts of nanosilver as produced on specific organisms, but in the real world, these studies don't accurately reflect the forms of silver that would be present if the produced nanosilver were released to the environment. Benn and Westerhoff (2007) and Geranio, Heuberger, and Nowack (2009) measured the release of nanosilver from nanosilver-treated socks in a laundering assay. Others have evaluated the fate of nanosilver in wastewater treatment systems (Kaegi et al. 2011; Kiser et al. 2010, 2012), demonstrating that some of the silver becomes soluble, that some of the particles settle to the solid sludge and are not released, and that some particles are released with the silver in the treated water, predominantly as silver sulfide nanoparticles. Thus, the nanosilver particles are mostly in the sludge (75%–90%), much of which is used as fertilizer (for growing, etc.), while some remains in the water and is released into water bodies. Some of these water bodies are the source water for public drinking water supplies.

Environmental chemistry is complex and varies greatly. In a saltwater environment, there will be a lot of mineral ions, including chlorides. In a wetland, there are a lot of sulfides and natural organic matter suspended in the water and deposited on the bottom (muck). As a ubiquitous component of aquatic systems, natural organic matter can influence the surface speciation and charge of nanoparticles, and thus affect their mobility and aggregation/deposition properties (Navarro et al. 2008). Natural organic matter may coat the surface of nanoparticles, resulting in particles that tend to stay dispersed rather than aggregate to larger particles and settle out (Handy, Owen, and Valsami-Jones 2008). Natural organic matter can stabilize particles against aggregation in water, which may enhance transport in aqueous environments and groundwater (Lowry and Casman 2009).

The presence of bacteria and other organisms also affects the fate of ENMs. For conventional chemical substances, their fate can be predicted by how water- or fat-soluble they are; for nanoparticles, the surface properties also contribute to fate. Kiser et al. (2010) found that functionalized nanoparticles were removed less during wastewater treatment than nonfunctionalized or coated nanoparticles. The presence of natural organic matter interfered with the biological interactions that normally remove substances during wastewater treatment. The U.S.EPA has proposed using a test method with freeze-dried sludge for testing under the New Chemicals Program in Toxic Substances Control Act (TSCA) for ENMs (U.S.EPA 2010b). Kiser et al. (2012) tested the removal of nine different ENMs with fresh versus freeze-dried activated wastewater-treatment sludge, and they found that the freeze-dried sludge was less effective at removing the ENMs, and that the test did not predict what actually happens during wastewater treatment. When heat drying was also applied, removal decreased further. The likely reason for this is that in activated sludge, bacteria bind the ENMs, aggregating them so they become large particles and settle out with the solids. The freeze drying and especially heat treatment kill many of the bacteria, and their contents are released into the solids. As others have shown, the presence of proteins and other molecules with ENMs results in the formation of a protein layer on the ENMs, which makes them more soluble and stable as nanoparticles. Kiser et al. (2012) describe this as behaving like surfactants. Thus, not only is U.S.EPA's proposed method inappropriate for measuring removal in wastewater treatment, any type of biomass treatment that results in heat or microbe-induced degradation, such as anaerobic digestion, composting, soils, or landfills, is likely to transform ENMs into more soluble particles.

The takeaway message is that while environmental chemistry predicts the formation of insoluble complexes of silver in natural waters, this assumption may not hold true for some novel nanosilver and other types of nanoscale particles engineered specifically to remain as small particles. Testing regimes must consider the surface properties, the presence of complexing agents, and the specific configuration of the ENMs to adequately predict their fate in the environment.

5.1.2 Titanium Dioxide (TiO$_2$)

One of the most widely studied nanomaterials for environmental effects is nanoscale titanium dioxide (nTiO$_2$). An interesting property of nTiO$_2$

is that it is photocatalytic—that is, it behaves as a catalyst when exposed to ultraviolet (UV) light. It has been explored for use in water treatment to destroy chemicals such as polychlorinated biphenyls (PCBs), pesticides, and other complex organic contaminants. Preparations of $nTiO_2$ have also been demonstrated to be bacteriocidal (Coleman et al. 2005; Kuhn et al. 2003; Rincon and Pulgarin 2003). Coleman et al. (2005) compared several $nTiO_2$ preparations for water treatment in a slurry and when immobilized. One commonly used type of $nTiO_2$ had a negative surface charge, resulting in an acidic pH in the water that created additional stress in *E. coli* bacteria (Coleman et al. 2005). Rincon and Pulgarin (2003) observed a detrimental effect on the survival of *E. coli* after photocatalytic exposure; no bacterial growth was observed after UV illumination of a contaminated $nTiO_2$ suspension. UV-illuminated $nTiO_2$ was more bactericidal to thin-walled gram-negative bacteria than thicker-walled gram-positive bacteria on surfaces, suggesting that the mechanism of toxicity is by radical hydroxyl generation on the cell membrane (Kuhn et al. 2003). Others have suggested lipid peroxidation as the mechanism of toxicity of photoilluminated $nTiO_2$ on cell membranes (Maness et al. 1999).

5.2 Estimating Environmental Exposures

As we've discussed, there are two components to risk: the nature of the hazard (toxicity) and the extent of contact with the hazard (exposure). In this section, we will consider some of the key issues related to environmental exposures to ENMs.

"As it turns out, nobody—no research institution, no government agency, no industry association—knows, even vaguely, how much nanomaterials are manufactured today" (Berger 2011). Surprising, but true. In 2009, the U.S.EPA compared their list of substances submitted under a voluntary program to the publicly available Consumer Products Inventory Project on Emerging Nanotechnology (WWCS 2011) and found only twelve overlapping products. The rest of the several hundred products were on one list and not the other. Frustrated by the lack of publicly available data, a group of academics tried to estimate the amount of five nanomaterials produced in the United States using Web searches, patents, and numbers of employees; the group produced low reliability estimates that are not even comparable to each other because different methods were used for each one (Hendren et al. 2011). In the United Kingdom, a group established an inventory of consumer products containing ENMs and estimated their environmental concentrations, suggesting inhalation exposures to nanoscale titanium dioxide and cerium oxide were on the order of 7 mg/m^3 and 7×10^{-7} mg/m^3, respectively (Boxall, Tiede, and Chaudhry 2008).

Another group estimated global production of nanoscale silver, titanium dioxide, and carbon nanotubes, and predicted their respective concentrations in Switzerland from diverse products (Mueller and Nowack 2008). This effort estimated atmospheric concentrations of about 0.001 micrograms per cubic meter (10^{-3} µg/m^3). To put this into more common terms, that's roughly a grain of salt's worth of ENMs per trillion parts of air, or about one nanoparticle inside a large refrigerator/freezer, or a small hot tub (a cubic meter is the volume of a box that is one meter tall, by one meter wide, by one meter deep). Estimates for soil were about 20 nanograms per kilogram of soil (20 parts per trillion) and, in fresh water, less than one part per trillion. Of course, the distribution of these particles won't be even. Higher concentrations might occur near wastewater treatment plants, highways, and industrial areas where ENM products might be released, or in parts of the food web.

5.3 Short-Term Toxicity Tests

One common measure of effects of substances on the environment is evaluating their toxicity to organisms in water. Several standardized tests measure the concentration of a substance required to kill or measurably decrease a population of test organisms, such as water fleas (daphnia) or small fish, like minnows. *Daphnia magna* is a species of water flea that is a filter feeder. They are small, transparent organisms that grow and reproduce quickly, and obtain nutrition by filtering water through their bodies. They are thus a simple organism to study and obtain results from, and so are used in several standardized assays (EPA, OECD, and EU) to evaluate short-term effects of substances in aquatic systems.

5.3.1 Daphnia LC$_{50}$ Assays

Standard assays for testing the aquatic toxicity of substances in daphnia measure the concentration that is lethal to 50% of the test population, the LC_{50}, or the effects concentration, EC_{50}. The lower the LC$_{50}$ concentration, the greater is the toxicity of a substance. For C$_{60}$ fullerenes, an LC$_{50}$ of 460 parts per billion (ppb) was found in 48-h toxicity tests with *D. magna* prepared with tetrahydrofuran (THF, a solvent) and 7.9 ppm for sonicated (using ultrasound to disperse) C$_{60}$, possibly indicating lower toxicity for the sonicated C$_{60}$. However, some difficulties were reported in particle dispersion in the sonicated experiments, so the results are not conclusive. Adult daphnids demonstrated behavioral irregularities when exposed to C_{60} (Lovern and Klaper 2006). In another study with *D. magna*, the results indicated uptake and sublethal effects of fullerene exposure, including altered molting and decreased reproductive output (Oberdörster et al. 2006). Daphnids exposed to single-walled carbon nanotubes (SWCNTs) coated with a lipid (fat) layer were able to metabolize the outer layer and excrete the SWCNTs back into the water (Roberts et al. 2007), demonstrating that biological organisms can modify nanoparticles in the environment.

Material preparation was also shown to affect the toxicity of $nTiO_2$. Using standard U.S.EPA 48-h acute toxicity tests in daphnids, Lovern and Klaper (2006) found toxicity associated with exposure to filtered tetrahydrofuran (THF)-derived 30-nm TiO_2 (classified as P25) particles and reported a 48-h lethal concentration (LC_{50}) of 5.4 ppm and a no-observed-effects concentration (NOEC) of 2.0 ppm. Unfiltered and sonicated TiO_2 particles that were agglomerated (stuck together to form larger particles) to 100 to 300 nm were less toxic to daphnids; mortality never exceeded 9%, and no LC_{50} value could be determined. This may have been due to differences in particle size, or perhaps the THF in the filtered experiments was the cause of toxicity rather than the TiO_2. There are some indications that daphnids can ingest nano-sized TiO_2, with particles in the gut and fatty lipid storage droplets appearing shortly after ingestion. It also appears that particles may be transported to other parts of the organism (Lovern and Klaper 2006).

In a modified standard bioassay in algae, Hund-Rinke and Simon (2006) demonstrated a difference in toxicity from UV-illuminated 25-nm TiO_2 (P25) and 100-nm particle diameter TiO_2. An effects concentration (EC_{50}) of 32 to 44 mg/L was found for the 25-nm diameter particles. According to the European Union (EU) Directive 67/548/EEC, this substance would be labeled harmful to aquatic organisms; however, under the U.S.EPA classification (U.S.EPA 2001), this result is classified as low acute toxicity. Similar to the fullerenes, for the 100-nm diameter particles, there was not enough toxicity to calculate an EC_{50}, although some toxicity was observed. Hund-Rinke and Simon (2006) compared washed particles to unwashed ones, and report slightly, but not significantly, lower toxicity with the washed particles. They further report no toxicity in experiments with daphnids; however, no measurements of particle aggregation or other properties were made. Wiench et al. (2007) compared the acute toxicity of nano- and microscale TiO_2 particles in daphnids using the OECD Test Guideline 202. Although the choice of test media affected the level of agglomeration and sedimentation, both particle sizes showed similarly low acute toxicity.

These findings of the toxicity of $nTiO_2$ can be compared to interpret the results (Lovern and Klaper 2006; Hund-Rinke and Simon 2006; Warheit et al. 2007; Wiench et al. 2007). Table 5.1 shows the toxicity levels in different

TABLE 5.1

Acute Toxicity of Nanoscale TiO$_2$ in Aquatic Tests

Test	Material	Study	Endpoint	Value	EPA Hazard Ranking
Acute aquatic invertebrate (daphnids)	THF 30 nm anatase	EPA 48 h tox test (Lovern and Klaper 2006)	LC$_{50}$ (48 h)	5.5 mg/L	M
Acute aquatic invertebrate (daphnids)	THF 30 nm anatase	EPA 48 h tox test (Lovern and Klaper 2006)	NOEC	2.0 mg/L	M
Acute aquatic invertebrate (daphnids)	Sonicated >100 nm anatase	EPA 48 h tox test (Lovern and Klaper 2006)	LC$_{50}$ (48 h)	NA	L
Acute aquatic invertebrate (daphnids)	25 nm P25 (80% anatase: 20% rutile)	EC standard algal assay (Hunde-Rinke and Simon 2006)	EC50	NA	L
Acute aquatic invertebrate (daphnids)	100 nm anatase (Hombikat)	EC standard algal assay (Hunde-Rinke and Simon 2006)	EC$_{50}$	NA	L
Acute aquatic invertebrate (daphnids)	Unknown	OECD 202 (Wiench et al. 2007)	EC50	NA	L
Acute aquatic invertebrate (daphnids)	140 nm 79% rutile: 21% anatase	OECD 202 (Warheit et al. 2007b)	EC$_{50}$ (48 h)	>100 mg/L	L
Acute algal toxicity	140 nm 79% rutile: 21% anatase	OECD 201 (Warheit et al. 2007b)	EC$_{50}$ (72-h growth)	21 ± 5 to 87 ± 4 mg/L	M
Acute algal toxicity	380 nm rutile	OECD 201 (Warheit et al. 2007b)	EC$_{50}$ (72-h growth)	16 ± 6 to 61 ± 9 mg/L	M
Acute fish toxicity test	140 nm 79% rutile: 21% anatase	OECD 203 (Warheit et al. 2007b)	LC^{50} (96 h)	>100 mg/L	L
Acute algal toxicity	25 nm P25 (80% anatase: 20% rutile)	EC standard algal assay (Hunde-Rinke and Simon 2006)	EC$_{50}$	32–44 mg/L	M (harmful according to EC)

(continued)

TABLE 5.1 (CONTINUED)

Acute Toxicity of Nanoscale TiO$_2$ in Aquatic Tests

Test	Material	Study	Endpoint	Value	EPA Hazard Ranking
Acute algal toxicity	100 nm anatase (Hombikat)	EC standard algal assay (Hunde-Rinke and Simon 2006)	EC$_{50}$	NA	L

Abbreviations: NA – not applicable; M – medium; L – low.

test systems as well as the hazard ranking according to the U.S.EPA Hazard Ranking Scale (U.S.EPA 2001). The nature of the particles and the tests vary, but generally report similar results in a low- to medium-hazard ranking for this substance.

Daphnids were exposed to nanoscale iron particles in water, and these were shown to be taken up into the digestive tract, but apparently not into the rest of the organism. The uptake did not affect their survival or reproduction, and nano-iron toxicity was similar to that found with larger iron particles. The only observed effects were that the antennae were clogged externally from the exposure and that their digestive tracts were darker, because the iron produced a dark color visible through the fairly transparent body of the small organisms (Oberdörster et al. 2006). These studies suggest that particle size may affect the toxicity of substances to aquatic organisms.

Templeton et al. (2006) tested the toxicity of SWCNTs in a saltwater organism, *Amphiascus tenuiremis* (also known as copepods), using a standard toxicity test (ASTM Method E-2317-04) and demonstrated that purified nanotubes were not toxic during a chronic (long-term) exposure. The SWCNTs were prepared by washing in nitric acid, which removed impurities. In contrast, the unpurified SWCNTs were toxic at the highest dose of 10 ppm, a higher concentration than would likely occur in the environment. A third experiment found that a smaller fraction of the produced material, which the authors called *fluorescent nanocarbon*, was also toxic to the copepods at 10 mg/L. This study demonstrated differences in the uptake of SWCNTs by size and purity. The larger purified nanotubes did not cross the gut, nor did they cause toxicity. The copepods excreted SWCNT as condensed clusters, transforming them into a new shape. The study also showed that impurities (i.e., the fluorescent nanocarbon) may be important contributors to toxicity and that organisms can transform nanomaterials in the environment (Templeton et al. 2006).

These studies indicate that material preparation affects the toxicity of nanomaterials and that test conditions also affect the results of toxicity testing. These initial results can be considered indicators, but not definitive findings. At this moment, the limited data only allow the conclusion that some nanoparticles may be toxic to small aquatic organisms. Much work is needed to understand what factors affect the study results and to produce reliable studies that can be considered more conclusive. According to the U.S.EPA, $nTiO_2$ would be of low to medium ecological toxicity in short-term tests.

5.4 Studies of Nanomaterial Toxicity to Fish

Relatively few studies have looked at the effects of nanomaterials on larger aquatic organisms such as fish. In this section, we discuss a few of the current studies.

5.4.1 Are Fullerenes Toxic to Fish?

Several studies with C_{60} fullerenes demonstrated that the preparation method affected absorption and toxicity. A 2004 study in largemouth bass found uptake of tetrahydrofuran (THF)-prepared C_{60} into the brain (Oberdörster 2004), but later studies suggest the effects observed may have been due to the solvent THF (Scown et al. 2010). Solvent-free preparations of C_{60} did not induce toxicity in fathead minnows or carp, but liver enzymes in goldfish were elevated following exposures well above environmental levels (Scown et al. 2010). While early reports stirred concerns about the toxicity of fullerenes to aquatic organisms, current views suggest the observations may be artifacts of experimental design. Importantly, a solvent mentioned previously, THF, which causes neurological, or brain, effects, was used in several early studies to help transfer the C_{60} particles into water because they are insoluble. It has been shown that C_{60} molecules form aggregates called nano-C_{60}, and nano-C_{60} does remain suspended in water, as if it were soluble. Several studies have now shown that using THF to solubilize C_{60} in water results in THF becoming part of the nano-C_{60}. Thus, the observed brain effects may have been a result of the THF exposure, not the fullerenes (Oberdörster, McClellan-Green, and Haasch 2006a; Henry, Petersen, and Compton 2011). A key issue is that fullerene molecules are not soluble in water. When fullerenes are substituted, for example with a hydroxyl group, they can become soluble, and toxicity resulting from reactive oxygen species is observed (Lee et al. 2007), albeit at low levels.

Oberdörster and colleagues (2006) also evaluated the effects of nano-C_{60} on the benthic (mud-dwelling) invertebrate *Hyalella azteca*, marine organisms, fathead minnows, and Japanese medaka (fish). As with TiO_2, LC_{50}

values in invertebrates could not be determined using the doses tested (exposure to nano-C_{60} resulted in less than 50% of the population dying). The studies in fish tell a complicated story of how the way experiments are designed affects the outcome.

As discussed previously, one of the main observations in toxicology studies is that introducing nanomaterials into biological systems generates reactive oxygen species (ROS). A review of earlier studies by Henry et al. (2011) suggests that experimental artifacts may be the cause of observed ROS generation, including solvent effects from THF, light that causes photoactivation of nano-C_{60}, and micellular preparations (above the critical micelle concentration). These authors cite seven studies associating ROS production with experimental factors, and conclude that "present evidence indicates that ROS production by aqueous nC_{60} is minimal."

One study that measured antioxidant effects in goldfish (*Crassius auratus*) reported significant changes in liver enzymes, glutathione, and levels of lipid peroxidase when exposed to nano-C_{60}. The results were not dose related: There were higher responses at the low versus the high dose (Zhu et al. 2008). These authors suggest the results might not be a direct effect of the nano-C_{60}, because growth was not inhibited at the low dose level. Generally, toxicologists expect to see greater effects at a higher concentration. Of course, aggregation can't be ruled out for the higher doses such that the measured dose rate does not reflect actual exposure levels. However, in the larger context of the weight of evidence, both Zhu et al. (2008) and Henry et al. (2011) conclude that experimental artifacts may be the cause of the observed effects and hence represent a false positive. However, a review of seventy-seven nano-ecotoxicological studies confirmed that, in general, existing assay methods are valid for nanomaterials and that their aquatic toxicity levels are similar to conventional substances (Kahru and Dubourguier 2010).

Still, these studies raise concerns about how nanomaterials are prepared, and how effects are measured in classical toxicology and ecotoxicology assays. Equally important is that they don't necessarily represent the form of the nanomaterial as it is likely to be present in the environment. Henry et al. (2011) state, "Perhaps solvent generated aqueous nC_{60} will become industrially important such that nC_{60} (solvent) exposure in fish becomes environmentally relevant." But we must more generally ask, in the real world—where there is complicated chemistry in aquatic systems, with the presence of natural organic matter, dissolved oxygen, many electrolytes, and especially if wastewater, runoff, or other contaminants are part of the system—what is the form of a nanomaterial that a fish or other aquatic species may be exposed to? Several studies shed light on this topic with regard to nC_{60} exposure and toxicity.

One investigation measured the effects of metals present in "as-produced" C_{60} and compared them with effects from waste generated by commercial production of C_{60} (Hull et al. 2009). The results suggest toxicity was due to the presence of metallic contaminants in C_{60} and raise concern about the findings from experiments with pure C_{60}, which may not be commercially produced. Another study measured effects of C_{60} on fish via dietary exposure instead of in the water. Low toxicity was observed, confirming other observations about experimental artifacts affecting toxicity (Fraser et al. 2011). Park et al. (2011) reported that nano-C_{60} associated with a cocontaminant (ethinyl estradiol, or EE) and reduced its bioavailability to zebra fish, but also reduced its biodegradation rate, making it potentially more persistent in the environment. This effect of nanoparticles binding to other molecules raises concerns about the Trojan horse effect, where absorption of a nanoparticle can bring along other contaminants and effectively increase exposure, as explained in the following discussion.

5.4.2 TiO$_2$ in Arsenic and Carp

It was recently demonstrated that arsenic strongly binds to nanoscale titanium dioxide (nTiO$_2$) in water. Further, the presence of nTiO$_2$ more than doubled the uptake of arsenic in carp (Sun et al. 2007). The nTiO$_2$ also accumulated in the fish and correlated with arsenic absorption. The presence of nTiO$_2$ did not alter the distribution of arsenic in the fish, however. Arsenic with nTiO$_2$ accumulated preferentially in the intestine, stomach, gills, liver, skin, and scales, and least in the muscle. While not quantitatively characterized in the study, the accumulation of nTiO$_2$ was much greater in the presence of arsenic compared to nTiO$_2$ alone. However, the fish were sacrificed to assess uptake, so no toxicity can be inferred from these experiments. These findings suggest a secondary environmental effect: increasing the uptake of other environmental contaminants and altering their environmental behavior.

These few studies indicate that while nanoparticles might be harmful to aquatic organisms, the way the studies were conducted affected the results. In some cases, the material preparation affected the measurements. As with the human toxicity studies, the impacts may be related to the smaller size of the particles, their increased reactivity as a result of greater surface area per particle, or the greater number of particles in a dose. Also as with human toxicity studies, new ways are needed to describe the doses of nanoparticles to allow more accurate interpretation of findings.

5.5 Natural Formation of Nanosilver in Water

A couple of recent studies shed light on the aquatic pathway for nanoparticles. It has long been known that silver's antimicrobial properties are due to the release of silver ions, which are toxic to bacteria. Studies of the environmental fate of nanosilver have suggested that most nanosilver binds to sulfur molecules naturally present in water, and forms insoluble complexes, recently determined to be mainly silver sulfide nanoparticles (Potera 2010; Kim et al. 2010). Two recent studies have shown that, in the presence of water or humidity, silver nanoparticles can form even when there were no silver nanoparticles to start with.

Akaighe et al. (2011) demonstrated that nanosilver forms in the presence of humic acids, water, and ultraviolet light. Noting that colloidal silver was first observed in natural water with silver ions present in the 1990s, before nanosilver was specifically manufactured (colloidal silver, which may have nanosize particles, was used then), the researchers sought to understand the dynamics of formation at different temperatures and with different types of humic acids, soluble complexes that naturally occur and contain natural organic matter, particularly in mucky rivers and lakes. They found that humic acids in soil, which have more aromatic sulfurs than sediment humic acids (which are more aliphatic), were less effective, suggesting that the organic sulfide (thiols) of the sediment humic acids is an important ingredient. The size distribution of the silver nanoparticles formed ranged from 5 to 50 nm in diameter, and most solutions included polydispersed sizes.

Another group found that placing silver nanoparticles on a surface (a silicon dioxide grid) in the presence of humidity resulted in daughter particles forming near the original ones as quickly as three hours after placing them there (Glover, Miller, and Hutchison 2011). The new nanoparticles formed more quickly in the presence of ultraviolet light (sunlight), which suggests they could form naturally. The astonishing finding was that not only did new particles form from existing nanoparticles, but from silverware and jewelry made of silver and copper!

These studies are significant and tell us that nanoparticles have been in our environment for a long time. Not only that, but we have probably been drinking and eating metallic nanoparticles for centuries! Silverware coated with—or made from—silver releases silver ions, and these form silver nanoparticles, and we eat them! The silver nanoparticles probably contributed to keeping people healthy prior to refrigeration and water treatment by killing pathogens in food and water. Copper pitchers and copper jewelry, both long associated with health in Eastern traditions, release copper ions, and these form copper nanoparticles. Without ever seeing them, we have traditionally made metallic products and been exposed to nanoparticles formed in nature by using them.

The significance of this finding cannot be overstated. There are many people very concerned about the use of nanosilver particles in consumer products. The fact that nanosilver has been in our environment for centuries should lessen that concern. The silver industry has been arguing a similar point for at least two years, that we have safely been using colloidal silver, which contains nanosilver particles, for over a century (Nowack et al. 2011).

That said, finding that two types of nanoparticles occur (and are created) in nature does not mean all nanoparticles are without harm. Some engineered nanoparticles are made specifically to harness novel properties and may be functionalized, have a unique shape, or comprise a novel combination of materials and properties that need to be studied. In fact, we still need to be mindful of the pathways of exposure to more common nanomaterials to ensure that they are safe when used in novel applications.

Further, this finding does not eliminate the need to understand how engineered nanosilver behaves as it is produced and used in consumer and other products. There may be differences in manufactured nanosilver that affect health and environmental risks differently from the more conventional sources of silver. However, it is clear that we have more history with some nanoparticles than previously realized.

5.6 Environmental Exposures

Many people argue that, unless nanoparticles are free particles, there is no need to worry about their effects in the environment. That is, when nanoparticles are embedded in a product matrix, there should be no concern about exposure to them because the material is no longer at the nanoscale: The particles cannot be absorbed because they are bound up in the product. However, there is evidence from other types of materials that embedded ingredients of products can be released to the environment.

For example, flame retardants are used in many consumer products. Clothing, chairs, tables, carpets, furniture, televisions, telephones and other electronics, and other products are coated with flame retardants, often required by law to limit their flammability and save lives. One category of flame retardants that has been widely used is polybrominated diphenyl ethers (PBDEs), which are not nanomaterials but are illustrative of potential environmental exposure concerns. Several recent studies measured the levels of PBDEs in dust in people's homes, airplanes, indoor air, and home electronics (Rudel et al. 2003; Stapleton et al. 2005; Wilford et al. 2005; EST Science News 2007a, 2007b). These studies demonstrate that there is widespread exposure to some PBDEs from consumer and electronic products. As products are used, small amounts of coating wear off and enter the environment. PBDEs have been measured in high levels in human blood in the U.S. population (Schecter et al. 2005); they have also been measured in polar bears—who are not sitting on couches treated with flame retardant fabrics, but are exposed because PBDEs were released to the environment and moved from North America and elsewhere to the Arctic regions. Many PBDEs are now banned, and it is possible in the foreseeable future that some nanomaterials could replace them as flame retardants. The point here is that it is necessary to test materials to see if they enter the environment, and not simply to assume that they will not.

5.6.1 Nanoscale Zerovalent Iron

Nanoscale zerovalent iron (NZVI) is being used at a number of hazardous waste sites to clean up chlorinated solvents that have contaminated groundwater. The metallic iron may be coated and is pumped into the ground, stimulating reactions including the breakdown of compounds that are in water below ground. NZVI is introduced in the groundwater to catalyze the removal of chlorine molecules from common solvents perchloroethylene and trichloroethylene (PCE and TCE) as well as PCBs and organochlorine pesticides (Karn, Kuiken, and Otto 2009). The dechlorination of these substances eventually breaks them down to carbon dioxide. Many hazardous waste sites are contaminated with these chlorinated solvents that were once widely used for cleaning, degreasing, and other purposes, and which persist in the groundwater. NZVI is accelerating the cleanup of these sites. NZVI has also been shown to immobilize arsenic, chromium, and lead, three highly toxic metals that have many industrial uses, which are also common at hazardous waste sites. In the cleanup process, as the NZVI particles oxidize and become rust particles, they reduce the contaminants in the groundwater to forms that are less mobile and less hazardous.

In air monitoring for ENM, background levels of ambient nanoscale particles pose difficulties for measurement. In the case of NZVI in groundwater, a difficulty is in detecting the NZVI in the presence of naturally occurring iron. Iron is prevalent in the environment, making up about 5% of the Earth's crust. However, iron generally occurs as minerals such as iron oxides and iron sulfides, not as pure metal. Zerovalent iron does not occur naturally. A body of research focuses on the factors affecting NZVI mobility, aggregation, and dispersion (e.g., Phenrat et al. 2010; Reinsch et al. 2010). As described in Chapter 6, Oberdörster et al. (2006) showed that nanoparticles of iron have similar toxicity to larger iron particles. However, there is so much naturally occurring iron in the soil, rock, and groundwater, it would be very difficult to detect nanoscale iron above these background levels. For arsenic removal, NZVI binds to the arsenic. This new material might have novel characteristics and could mobilize the arsenic in the subsurface environment.

Many researchers are focused on improving the efficacy of NZVI remediation. For example, the effectiveness of the NZVI is diminished by aging of particles, even during shipping and storage. NZVI is reactive, and as it oxidizes, it become less effective, binds other molecules, and if not coated, easily forms aggregates that lower efficacy and slow migration. At lower concentrations, there is less aggregation, and the individual nanoparticles last longer. As larger molecules, NZVI particles cannot travel far in groundwater and become stuck between soil particles; they are also less reactive because the surfaces are stuck to each other, and thus cannot react with and reduce the chlorinated solvents.

These data suggest that NZVI is not yet traveling far in the groundwater. The implication is that researchers are now seeking ways to enhance the mobility of NZVI so that particles will last longer as nanoparticles, continuing to reduce the solvents, and will travel further from the point of origin to clean up more pollution. As already discussed for other types of ENM, the form, surface characteristics, and environmental factors affect the toxicity of NZVI. The toxicity of NZVI to cells, bacteria, and fish relates to the reactivity of the surface that creates reactive oxygen species (Phenrat et al. 2009; Auffan et al. 2008; Li et al. 2010; Cullen et al. 2011; Chen et al. 2011). As the performance of NZVI is enhanced, its potential to migrate in the environment also increases, and the behavior of the enhanced materials needs to be evaluated in the environments it reaches.

5.7 Risk Assessments

Having surveyed some of the toxicity and exposure data for nanoparticles, let us now consider how these data can be used for risk assessment.

5.7.1 NIOSH—TiO$_2$

The National Institute for Occupational Safety and Health (NIOSH 2005) recently reviewed animal and epidemiological studies of TiO$_2$, but these primarily pertained to pigment-grade TiO$_2$, as manufacturing of ultrafine TiO$_2$ did not start until the 1990s. The draft document (NIOSH 2005) stated that tumorigenic effects of TiO$_2$ in rats "appear[ed] to be a function of particle size and surface area acting through a secondary genotoxic mechanism associated with persistent inflammation." The draft indicated that insufficient evidence exists to designate TiO$_2$ as a potential occupational carcinogen (cancer-causing substance), but noted concern about the potential carcinogenicity of ultrafine (nanoscale) TiO$_2$ if exposure levels are at the current occupational limits. According to the draft, studies in rats demonstrated a dose–response relationship for pulmonary effects when the dose of TiO$_2$ was expressed on the basis of surface area. Based on this information, a draft exposure limit of 1.5 mg/m^3 for fine (<2.5 μm) and 0.1 mg/m^3 for ultrafine (<0.1 μm, i.e., <100 nm) TiO_2 was recommended (NIOSH 2005). These estimates have been revised since, based on new evidence, and as discussed in Chapter 8, NIOSH currently recommends an exposure limit for ultrafine TiO_2 of 0.3 mg/m^3 in the workplace (NIOSH 2005).

The NIOSH REL (recommended exposure limit) is lower than the Japanese recommended Occupational Exposure Limit of 0.61 mg/m^3 (Nakanishi 2011), which is based on one specific type of TiO_2 nanoparticle called P25. In their analysis, Nakanishi and colleagues determined that other types of TiO_2 nanoparticles caused less inflammatory activity than P25, so that the limit would be applicable for a wide variety of TiO_2 nanoparticles. In their analysis, Nakanishi summarizes the findings of ten studies of "nano-materials toxicity tests," showing that "surprisingly few test samples having a primarily particle size (average value) less than 100 nm, and that the secondary particle size (average value) of almost all the test samples ranges from a few hundred nm to over 1 µm" (Nakanishi 2011).

In the risk assessment, Nakanishi reports that exposures in the workplace were generally to aggregated particles of at least several hundred nanoparticles, up to a few microns in diameter. The basis of the risk assessment was inflammation from the inhaled particles, a physical effect of the particles, and not a chemical effect, at a No Observable Adverse Effect Level (NOAEL) of about 2 (1.82 human equivalent) mg/m^3, an uncertainty factor of 3 was applied to derive the value of 0.61 mg/m^3 (Nakanishi 2011).

5.7.2 Carbon Nanotubes

Several organizations have conducted risk assessments of exposure to carbon nanotubes (CNT), including the U.S. National Institute for Occupational Safety and Health (NIOSH 2011), the Japanese New Energy and Industrial Technology Development Organization (NEDO; Nakanishi 2011), and two industrial producers of carbon nanotubes, BASF and Bayer (Ma-Hock et al. 2009; BMS 2008). Each of these assessments focused on studies of one type of carbon nanotube and considered inhalation exposure in an occupational environment to raw nanotubes. The resulting occupational exposure limits range from 0.0025 mg/m^3 (BASF) to 0.03 mg/m^3 (Nakanishi 2011). In the review of existing studies and derivations, the differences are highlighted as they relate to the metrics of exposure, that is, that if compared on the basis of specific surface area rather than mass, the levels associated with inflammation are similar for the different materials. As described in Chapter 4, studies suggest that different properties of CNTs affect their toxicity, including length, purity, number of walls, etc.

Nakanishi (2011) shows that estimated exposure levels are low compared to these limits. Further, in Japan, the recommendation is for a ten-year exposure limit, assuming that the limits will be reestablished in that time frame, given the emerging nature of our current understanding of exposure, toxicity and risk. This idea of an adaptive approach to managing risks is discussed in greater detail in Chapter 6 as a logical approach to managing uncertainty for emerging risks, in which levels are established to be protective based on current knowledge, but that are revisited when new information is developed.

Given the low exposure levels to raw materials in the workplace, and the relative ease of measuring and limiting exposure to these materials in a controlled environment, several groups have focused on risks associated with the potential for release of CNT from polymers at later stages of the product life cycle (Kohler et al. 2008; Petersen et al. 2011; Wohlleben et al. 2011; Nguyen et al. 2011). These studies consider what might happen to carbon nanotubes in products such as polymers or fabrics when they are worn, abraded (by sanding, cutting, or other mechanical stress), burned, or degraded by weathering (sunlight and water). Petersen et al. (2011) reviewed 180 studies and reported on the state of evidence for a variety of pathways. Cena and Peters (2011) describe what is measured from specific products in an experimental setup, so-called lollipops of CNT stuck to pieces of polymer (further described by principal investigator Thomas Peters in Chapter 8). Generally, these studies suggest that some processes may degrade the polymer and create the potential for exposure to CNT from products containing them, particularly in the situation of weathering, where the CNT become brittle and concentrate on the surface, where they could be abraded. These exposures will be lower than during production stages, but they could occur under less controlled situations and with a wider variety of people and environmental "receptors."

For example, in their review, Petersen et al. (2011) suggest that while the studies showed degradation under ultraviolet radiation exposure, studies to look at releases during fires found masses of CNT accumulating near the surface that were not released into the air (but could be if touched!). The burning studies did not find CNT in particles in air, and in further simulations of waste incineration, the data suggest they would end up in ash, not released into the air.

Another potential pathway of environmental exposure is through wastewater treatment. If CNTs are released from fabrics or other products into water, then it is important to understand what happens to (the fate of) CNT in wastewater treatment. The short answer is, it depends on the type of CNT, whether it is functionalized, what type of water it is in, what else is in the water, and the types of chemicals used in the wastewater treatment. Petersen et al. (2011) reviewed several studies of model wastewater treatment systems, and the results show that whether CNT remain soluble in the water column, or are removed by solids, depends on the surface of the nanotubes, their concentration, and what is in the water for them to react with.

A recent study of MWCNT demonstrated that natural organic matter (NOM) can stabilize a generally insoluble substance in water. Hyung et al. (2007) demonstrated that laboratory solutions of NOM, as well as actual river water containing NOM, kept the normally insoluble MWCNT in solution for four months. Using an instrument called a Thermal Optical Transmittance Analyzer, this group was able to quantitatively measure the concentration of the MWCNT in solution, and showed that with additional NOM, increasing amounts of MWCNT were suspended as individual nanotubes.

Since most natural waters contain some NOM, this is an important finding. It suggests that under the right conditions, CNTs can be dispersed, rather than bundled; thus, they would be at the size that aquatic organisms could ingest them with food, and they potentially could travel long distances in water bodies. Previously, it was thought that nanomaterials would aggregate in water, as has been shown for nanoparticles in air. Recall, however, that there are many different types of nanotubes, and there are an equally diverse number of ambient water conditions.

In my opinion, despite the finding that dissolved organic matter can stabilize CNTs in water, CNTs are generally insoluble and are bound in products, and thus are unlikely to occur as nanoparticles in ambient waters. Even in the unlikely case where free CNTs are released into water, the surface of most CNTs is generally very reactive, and they will likely bind to a broad range of biological and other particles in the water and rapidly accumulate into much larger particles, and settle into the sediment. Petersen et al. (2011) also suggest there is a low likelihood of wastewater treatment being a significant exposure pathway for CNTs. That said, if CNTs did enter wastewater treatment, they are highly toxic to bacteria and viruses, and would probably cause cytotoxicity. In fact, CNTs are being investigated as a water treatment agent because of their reactivity. The few existing studies suggest that the nature and amount of starting material as well as the water quality and methods of wastewater treatment all affect removal rates of CNTs in simulated treatment, which varied from 10% to 85% removal in one simulation and 96.3% to 99.7% in others. If the starting material is functionalized, it may be more difficult to remove by treatment because "raw" CNTs are highly reactive and would be anticipated to bind to other matter and settle quickly.

As discussed previously, studies of the mobility of CNTs suggest, again, that concentration, interaction chemistry (CNT surface and media), and velocity are critical components. If functionalized so they remain soluble, CNTs can move very quickly, but they can have physical barriers, owing to their rodlike shape. Smaller tubes would be expected to have greater mobility. Studies have shown that the presence of dissolved organic matter (muck) can bind CNTs and keep them soluble in the water column. A few studies suggest that in salty or hard water,

CNTs may settle out and bind to sediments. But, the conclusion is the same as in wastewater treatment. The fate of CNTs in soils and water environments will depend on the surface chemistry of the CNTs, the presence of ions and molecules, and factors such as the rate of flow and concentration that affect the ability of CNTs to bind other molecules, including other CNTs, and particles. There is a need to understand the complex dynamics of these reactive particulates and the factors that affect their behavior.

One group I am involved in is currently working to develop standardized methods for measuring the release of nanoparticles from solid matrices (Nanorelease project). Different from more traditional chemical substances, it is difficult to measure CNTs, especially in the presence of other carbon compounds. Presently, visual methods are needed to confirm particle measurements, and it can be difficult to "see" CNTs when other matter is present. One such group is working toward methods that will measure and report the amount of CNTs released under specific conditions, using a variety of types of starting material. The Nanorelease project is a coalition of governmental, nongovernmental, industry, and academic scientists who recognize the need to have a way to measure different materials using similar methods. The Nanorelease project is gathering experts to define, obtain, and characterize the starting materials, develop an initial array of measurement techniques, and create a round-robin testing scheme in different laboratories to assess the reliability and accuracy of methods. This ongoing work will advance the science and allow the comparison across products, something very difficult to attempt when different methods and materials are used in reported studies (ILSI 2012).

5.8 Summary and Conclusion

The complexity of issues to address in risk assessment for the broad class of nanomaterials and nanotechnologies currently in development requires a comprehensive approach. Current frameworks need to consider both the factors contributing to human health effects that need to be evaluated as well as ecological effects; and not simply the raw material, but the material as it is used and as it may be transformed in the environment. This requires a detailed understanding of the behavior of materials in the environment. It requires adopting a life cycle approach to risk analysis, and it requires more focus on exposure assessment early in the evaluation process. Chapters 6 and 7 describe NANO LCRA (life cycle risk assessment) and several other proposed frameworks for assessing and managing nanoscale materials across their life cycles.

References

Akaighe, N., R. I. MacCuspie, D. A. Navarro, D. S. Aga, S. Banerjee, M. Sohn, and V. K. Sharma. 2011. Humic acid–induced silver nanoparticle formation under environmentally relevant conditions. *Environ. Sci. Technol.* 45(9): 3895–901.

Asharani, P. V., Y. L. Wu, Z. Gong, and S. Valiyaveettil. 2008. Toxicity of silver nanoparticles in zebrafish models. *Nanotechnology* 19(25): 255102.

Auffan, M., W. Achouak, J. Rose, M. A. Roncato, C. Chanéac, D. T. Waite, A. Masion, J. C. Woicik, M. R. Wiesner, and J. Y. Bottero. 2008. Relation between the redox state of iron-based nanoparticles and their cytotoxicity toward *Escherichia coli*. *Environ. Sci. Technol.* 42(17): 6730–35.

Benn, T. M., and P. Westerhoff. 2008. Nanoparticle silver released into water from commercially available sock fabrics. *Environ. Sci. Technol.* 42(11): 4133–39.

BMS. 2008. Occupational exposure limit (OEL) for Baytubes defined by Bayer MaterialScience. http://www.baytubes.com/news_and_services/news_091126_oel.html

Boxall, A. B. A., K. Tiede, and Q. Chaudhry. 2007. Engineered nanomaterials in soils and water: How do they behave and could they pose a risk to human health? *Nanomedicine* 2(6): 919–27.

Berger, M. 2011. Does anyone know how much nanomaterials are produced? Anyone...? *Nanowerk*, April 11. http://www.nanowerk.com/spotlight/spotid=20942.php

Cena, L. G., and T. M. Peters. 2011. Characterization and control of airborne parti¬cles emitted during production of epoxy/carbon nanotube nanocomposites. *J. Occup. Environ. Hyg.* 8(2): 86–92.

Chen, P. J., C. H. Su, C. Y. Tseng, S. W. Tan, and C. H. Cheng. 2011. Toxicity assessments of nanoscale zerovalent iron and its oxidation products in medaka (*Oryzias latipes*) fish. *Mar. Pollut. Bull.* 63(5–12): 339–46.

Coleman, H. M., C. P. Marquis, J. A. Scott, S.-S. Chin, and R. Amal. 2005. Bactericidal effects of titanium dioxide-based photocatalysts. *Chem. Eng. Journal.* 113:55–63.

Cullen, L. G., E. L. Tilston, G. R. Mitchell, C. D. Collins, and L. J. Shaw. 2011. Assessing the impact of nano- and micro-scale zerovalent iron particles on soil microbial activities: Particle reactivity interferes with assay conditions and interpretation of genuine microbial effects. *Chemosphere* 82(11): 1675–82.

EPA. 2001. Printed wire board (PWB) surface finishes—Cleaner technologies substitutes assessment—Final. EPA/744-R-01-003A, Appendices B and H. http://www.epa.gov/opptintr/dfe/pubs/pwb/ctsasurf/pwb-pub.htm

————. 2010a. Interim technical guidance for assessing screening level environmental fate and transport of, and general population, consumer, and environmental exposure to nanomaterials. http://www.epa.gov/oppt/exposure/pubs/nanomaterial.pdf

————. 2010b. Nanomaterial case study: Nanoscale silver in disinfectant spray (external review draft). EPA/600/R-10/081. Washington, DC: U.S. Environmental Protection Agency.

EST Science News. 2007a. Flying high with PBDEs. DOI: 10.1021/es072571y

————. 2007b. Finding PBDEs in couches and TVs. http://pubs.acs.org/subscribe/journals/esthag-w/2007/may/science/kb_pbde.html

FDA. 1999. FDA bans colloidal silver products, cites lack of data. *FDA Consumer* 33(6): 5.

Fraser, T. W., H. C. Reinardy, B. J. Shaw, T. B. Henry, and R. D. Handy. 2011. Dietary toxicity of single-walled carbon nanotubes and fullerenes (C_{60}) in rainbow trout (*Oncorhynchus mykiss*). *Nanotoxicology* 5(1): 98–108.

Geranio, L., M. Heuberger, and B. Nowack. 2009. Behavior of silver nanotextiles during washing. *Environ. Sci. Technol.* 2009, 43:8113–18.

Glover R. D., J. M. Miller, and J. E. Hutchison. 2011. Generation of metal nanoparticles from silver and copper objects: Nanoparticle dynamics on surfaces and potential sources of nanoparticles in the environment. *ACS Nano* 5(11): 8950–57.

Handy, R. D., E. R. Owen, and E. Valsami-Jones. 2008. The ecotoxicology of nanoparticles and nanomaterials: Current status, knowledge gaps, challenges, and future needs. *Ecotoxicology* 17:315–25.

Hendren, C. O., X. Mesnard, J. Droge, and M. R. Wiesner. 2011. Production data for five engineered nanomaterials as a basis for exposure assessment. *Environ. Sci. Technol.* 45(7): 2562–69.

Henry, T. B., E. J. Petersen, and R. N. Compton. 2011. Aqueous fullerene aggregates (nC_{60}) generate minimal reactive oxygen species and are of low toxicity in fish: A revision of previous reports. *Curr. Opinion Biotech.* 22:533–37.

Hull, M. S., A. J. Kennedy, J. A. Steevens, A. J. Bednar, C. A. Weiss Jr, and P. J. Vikesland. 2009. Release of metal impurities from carbon nanomaterials influences aquatic toxicity. *Environ. Sci. Technol.* 43(11): 4169–74.

Hund-Rinke, K., and M. Simon. 2006. Ecotoxicological effects of photocatalytic active nanoparticles TiO2 on algae and daphnids. *Environ. Sci. & Pollut. Res.* 13(4): 225–32.

Hyung, H., J. D. Fortner, J. B. Hughes, and J. H. Kim. 2007. Natural organic matter stabilizes carbon nanotubes in the aqueous phase. *Environ. Sci. Technol.* 41(1): 179–84.

ILSA. 2012. Nanorelease project. http://www.ilsi.org/ResearchFoundation/Pages/NanoRelease1.aspx

Jacobs, M. N., G. T. Nolan, and S. R. Hood. 2005. Lignins, bactericides and organochlorine compounds activate the human pregnane X receptor (PXR). *Toxicol. App. Pharmacol.* 209(2): 123–33.

Kaegi, R., A. Voegelin, B. Sinnet, S. Zuleeg, H. Hagendorfer, M. Burkhardt, and H. Siegrist. 2011. Behavior of metallic silver nanoparticles in a pilot wastewater treatment plant. *Environ. Sci. Technol.* 45(9): 3902–8.

Kahru, A., and H. C. Dubourguier. 2010. From ecotoxicology to nanoecotoxicology. *Toxicology* 269(2–3): 105–19.

Karn, B., T. Kuiken, and M. Otto. 2009. Nanotechnology and in situ remediation: A review of the benefits and potential risks. *Environ. Health Perspect.* 117(12): 1813–31.

Kim, B., C. S. Park, M. Murayama, and M. F. Hochella. 2010. Discovery and characterization of silver sulfide nanoparticles in final sewage sludge products. *Environ. Sci. Technol.* 44(19): 7509–14.

Kiser, M. A., D. A. Ladner, K. D. Hristovski, and P. K. Westerhoff. 2012. Nanomaterial transformation and association with fresh and freeze-dried wastewater activated sludge: Implications for testing protocol and environmental fate. *Environ. Sci. Technol.* Published online first, doi:10.1021/es300339x

Kiser, M. A., H. Ryu, H. Jang, K. Hristovski, and P. Westerhoff. 2010. Biosorption of nanoparticles to heterotrophic wastewater biomass. *Water Res.* 44(14): 4105–14.

Köhler, A., C. Som, A. Helland, and F. Gottschalk. 2008. Studying the potential release of carbon nanotubes throughout the application life cycle. *J. Cleaner Production* 16:927–37.

Kuhn, K. P., I. F. Chaberny, K. Massholder, M. Stickler, V. W. Benz, H. G. Sonntag, and L. Erdinger. 2003. Disinfection of surfaces by photocatalytic oxidation with titanium dioxide and UVA light. *Chemosphere* 53(1): 71–77.

Lee, M. K., S. J. Lim, and C. K. Kim. 2007. Preparation, characterization and in vitro cytotoxicity of paclitaxel-loaded sterically stabilized solid lipid nanoparticles. *Biomaterials* 28(12): 2137–46.

Li, Z., K. Greden, P. J. Alvarez, K. B. Gregory, and G. V. Lowry. 2010. Adsorbed polymer and NOM limits adhesion and toxicity of nano scale zerovalent iron to *E. coli. Environ. Sci. Technol.* 44(9): 3462–67.

Lovern, S. B., and R. Klaper. 2006. *Daphnia magna* mortality when exposed to titanium dioxide and fullerene (C_{60}) nanoparticles. *Environ. Toxicol. Chem.* 25(4): 1132–37.

Lowry, G. V., and E. Casman. 2009. Nanomaterial transport, transformation, and fate in the environment: A risk-based perspective on research needs. In *Risk, Uncertainty and Decision Analysis for Nanomaterials: Environmental Risks and Benefits and Emerging Consumer Products*, ed. I. Linkov and J. Stevens, 125–139. Berlin: Springer Verlag.

Ma-Hock, L., S. Treumann, V. Strauss, S. Brill, F. Luizi, M. Mertler, K. Wiench, A. O. Gamer, B. van Ravenzwaay, and R. Landsiedel. 2009. Inhalation toxicity of multiwall carbon nanotubes in rats exposed for 3 months. *Toxicol. Sci.* 112(2): 468–81.

Maness, P.-C., S. Smolinski, D. M. Blake, Z. Huang, E. J. Wolfrum, and W. A. Jacoby. 1999. Bactericidal activity of photocatalytic TiO2 reaction: Toward an understanding of its killing mechanism. *Appl. Environ. Microbiol.* 65(9): 4094–98.

Mueller, N. C., and B. Nowack. 2008. Exposure modeling of engineered nanoparticles in the environment. *Environ. Sci. Technol.* 42(12): 4447–53.

Nakanishi, J. 2011. Risk assessment of manufactured nanomaterials final report: Overview of approaches and results and executive summaries of three manufactured nanomaterials: carbon nanotubes (CNT), fullerene (C_{60}), and titanium dioxide. Kawasaki: New Energy and Industrial Technology Development Organization of Japan.

Navarro, E., F. Piccapietra, B. Wagner, F. Marconi, R. Kaegi, N. Odzak, L. Sigg, and R. Behra. 2008. Toxicity of silver nanoparticles to *Chlamydomonas reinhardtii*. *Environ. Sci. Technol.* 42(23): 8959–64.

Nguyen, T., B. Pellegrin, C. Bernard, X. Gu, J. M. Gorham, P. Stutzman, D. Stanley, et al. 2011. Fate of nanoparticles during life cycle of polymer nanocomposites. *J. Phys.: Conf. Ser.* 304: 012060.

NIOSH. 2005. Draft NIOSH current intelligence bulletin: Evaluation of health hazard and recommendations for occupational exposure to titanium dioxide. www.cdc.gov/niosh/updates/upd-11-23-05.html

NIOSH. March 2011. NIOSH Docket Number 161A. Draft Current Intelligence Bulletin: Occupational Exposure to Carbon Nanotubes and Nanofibers. www.cdc.gov/niosh/docket/archive/docket161A.html

Nowack, B., H. F. Krug, and M. Height. 2011. 120 years of nanosilver history: Implications for policy makers. *Environ. Sci. Technol.* 45:1177–83.

NTP (National Toxicology Program). 2007. Department of Health and Human Services National Toxicology Program (NTP) Office of Chemical Nomination and Selection; announcement of and request for public comment on toxicological study nominations to the NTP. *Federal Register*, March 29. 72(60): 14816.

Oberdörster, E. 2004. Manufactured nanomaterials (fullerenes, C_{60}) induce oxidative stress in the brain of juvenile largemouth bass. *Environ. Health Perspect.* 112(10): 1058–62.

Oberdörster, G, A. Maynard, K. Donaldson, V. Castranova, J. Fitzpatrick, K. Ausman, J. Carter, et al. 2005. Principles for characterizing the potential human health effects from exposure to nanomaterials: Elements of a screening strategy. *Particle Fibre Toxicol.* 2:8.

Pal, S., Y. K. Tak, and J. M. Song. 2007. Does the antibacterial activity of silver nanoparticles depend on the shape of the nanoparticle? A study of the gram-negative bacterium *Escherichia coli*. *Appl. Environ. Microbiol.* 73(6): 1712–20.

Park, J. W., T. B. Henry, S. Ard, F. M. Menn, R. N. Compton, and G. S. Sayler. 2011. The association between nC_{60} and 17α-ethinylestradiol (EE2) decreases EE2 bioavailability in zebrafish and alters nanoaggregate characteristics. *Nanotoxicology* 5(3): 406–16.

Petersen, E. J., Zhang, L., Mattison, N. T., O'Carroll, D. M., Whelton, A. J., Uddin, N., Nguyen, T., Huang, Q., Henry, T. B., Holbrook, R. D., Chen, K. L. 2011. Potential release pathways, environmental fate, and ecological risks of carbon nanotubes. *Environ Sci Technol.* 45(23): 9837–56.

Phenrat, T., H. J. Kim, F. Fagerlund, T. Illangasekare, and G. V. Lowry. 2010. Empirical correlations to estimate agglomerate size and deposition during injection of a polyelectrolyte-modified FeO nanoparticle at high particle concentration in saturated sand. *J. Contam. Hydrol.* 118(3–4): 152–64.

Phenrat, T., T. C. Long, G. V. Lowry, and B. Veronesi. 2009. Partial oxidation ("aging") and surface modification decrease the toxicity of nanosized zerovalent iron. *Environ. Sci. Technol.* 43(1): 195–200.

Potera, C. 2010. Transformation of silver nanoparticles in sewage sludge. *Environ. Health Perspect.* 118(12): A526–27.

Pruden, A., R. Pei, H. Storteboom, and K. H. Carlson. 2006. Antibiotic resistance gene as emerging contaminants: Studies in northern Colorado. *Environ. Sci. Tech.* 40(23): 7445–50.

Reinsch, B. C., B. Forsberg, R. L. Penn, C. S. Kim, and G. V. Lowry. 2010. Chemical transformations during aging of zerovalent iron nanoparticles in the presence of common groundwater dissolved constituents. *Environ. Sci. Technol.* 44(9): 3455–61.

Rincon, A. G., and C. Pulgarin. 2003. Photocatalytic inactivation of *E. coli*: Effect of (continuous-intermittent) light intensity and of (suspended-fixed) TiO_2 concentration. *Applied Catalysis B—Environmental* 44: 263–84.

Roberts A. P., A. S. Mount, B. Seda, J. Souther, R. Qiao, S. Lin, P. C. Ke, A. M. Rao, and S. J. Klaine. 2007. In vivo biomodification of lipid-coated carbon nanotubes in *Daphnia magna*. *Environ. Sci. Technol.* 41: 3025–29.

Rudel, R. A., D. E. Camann, J. D. Spengler, L. R. Korn, and J. G. Brody. 2003. Phthalates, alkylphenols, pesticides, polybrominated diphenyl ethers, and other endocrine-disrupting compounds in indoor air and dust. *Environ. Sci. Technol.* 37(20): 4543–53.

Schecter, A., O. Papke, K. C. Tung, J. J. Harris, and T. R. Dahlgren. 2005. Polybrominated diphenyl ether flame retardants in the U.S. population: Current levels, temporal trends, and comparison with dioxins, dibenzofurans, and polychlorinated biphenyls. *J. Occup. Environ. Med.* 47(3): 199–211.

Scown, T. M., R. van Aerle, and C. R. Tyler. 2010. Review: Do engineered nanoparticles pose a significant threat to the aquatic environment? *Crit. Rev. Toxicol.* 40(7): 653–670.

Sondi, I., and B. Salopek-Sondi. 2004. Silver nanoparticles as antimicrobial agent: A case study on *E. coli* as a model for gram-negative bacteria. *J. Colloid Interface Sci.* 275(1): 177–82.

Stapleton, H. M., N. G. Dodder, J. H. Offenberg, M. M. Schantz, and S. A. Wise. 2005. Polybrominated diphenyl ethers in house dust and clothes dryer lint. *Environ. Sci. Technol.* 39(4): 925–31.

Sun, H., X. Zhang, O. Niu, Y. Chen, and J. C. Crittenden. 2007. Enhanced accumulation of arsenate in carp in the presence of titanium dioxide nanoparticles. *Water Air Soil Pollut.* 178:245–54.

Templeton, R. C., P. L. Ferguson, K. M. Washburn, W. A. Scrivens, and G. T. Chandler. 2006. Life-cycle effects of single-walled carbon nanotubes (SWNTs) on an estuarine meiobenthic copepod. *Environ. Sci. Technol.* 40(23): 7387–93.

Usenko, C. Y., S. L. Harper, and R. L. Tanguay. 2007. In vivo evaluation of carbon fullerene toxicity using embryonic zebrafish. *Carbon* 45(9): 1891–98.

Veldhoen, N., R. C. Skirrow, H. Osachoff, H. Wigmore, D. J. Clapson, M. P. Gunderson, G. Van Aggelen, and C. C. Helbing. 2006. The bactericidal agent triclosan modulates thyroid hormone-associated gene expression and disrupts post-embryonic anuran development. *Aquatic Toxicol.* 80(3): 217–27.

Warheit, D. B., R. A. Hoke, C. Finlay, E. M. Donner, K. L. Reed, and C. M. Sayes. 2007. Development of a base set of toxicity tests using ultrafine TiO_2 particles as a component of nanoparticle risk management. *Toxicol. Lett.* 171(3): 99–11C.

Wiench, K., R. Landsiedel, S. Zok, V. Hisgen, K. Radkel, and B. van Ravenzwaay. 2007. Aquatic fate and toxicity of nanoparticles: Agglomeration, sedimentation and effects on *Daphnia magna*. Abstract no. 1384:2007 Itinerary Planner. Charlotte, NC: Society of Toxicology.

Wilford B. H., M. Shoeib, T. Harner, J. Zhu, and K. C. Jones. 2005. Polybrominated diphenyl ethers in indoor dust in Ottawa, Canada: Implications for sources and exposure. *Environ. Sci. Technol.* 39(18): 7027–35.

Wohlleben, W., S. Brill, M. W. Meier, M. Mertler, G. Cox, S. Hirth, B. von Vacano, et al. 2011. On the lifecycle of nanocomposites: Comparing released fragments and their in vivo hazards from three release mechanisms and four nanocomposites. *Small: Nano Micro* 7(16): 2384–95.

WWCS (Woodrow Wilson International Center for Scholars). 2011. Project on emerging nanotechnologies: A nanotechnology consumer products inventory. http://www.nanotechproject.org/44

Zhu, X., L. Zhu, Y. Lang, and Y. Chen. 2008. Oxidative stress and growth inhibition in the freshwater fish *Carassius auratus* induced by chronic exposure to sublethal fullerene aggregates. *Environ. Toxicol. Chem.* 27(9): 1979–85.

Zhu, X., L. Zhu, Y. Li, Z. Duan, W. Chen, and P. J. Alvarez. 2007. Developmental toxicity in zebrafish (*Danio rerio*) embryos after exposure to manufactured nanomaterials: Buckminsterfullerene aggregates (nC$_{60}$) and fullerol. *Environ. Toxicol. Chem.* 26(5): 976–79.

Further Reading

CDC. 2005. *Third national report on human exposure to environmental chemicals.* Atlanta, GA: Centers for Disease Control.

Fortner, J. D., D. Y. Lyon, C. M. Sayes, A. M. Boyd, J. C. Falkner, E. M. Hotze, L. B. Alemany, et al. 2005. C$_{60}$ in water: Nanocrystal formation and microbial response. *Environ. Sci. Technol.* 39(11): 4307–16.

Health Effects Institute. 2001. Evaluation of human health risk from cerium added to diesel fuel. HEI Communication 9. http://pubs.healtheffects.org/view.php?id=172

Liu, Y., and G. V. Lowry. 2006. Effect of particle age (FeO) and solution pH on NZVI reactivity: H2 evolution and TCE dechlorination. *Environ. Sci. Technol.* 40(19): 6085–90.

Liu, Y., S. A. Majetich, R. D. Tilton, D. S. Sholl, and G. V. Lowry. 2005. TCE dechlorination rates, pathways, and efficiency of nanoscale iron particles with different properties. *Environ. Sci. Technol.* 39(5): 1338–45.

Lok, C. N., C. M. Ho, R. Chen, Q. Y. He, W. Y. Yu, H. Sun, P. K. Tam, J. F. Chiu, and C. M. Che. 2006. Proteomic analysis of the mode of antibacterial action of silver nanoparticles. *J. Proteome Res.* 5(4): 916–24.

Lowry, G. V., and K. M. Johnson. 2004. Congener-specific dechlorination of dissolved PCBs by microscale and nanoscale zerovalent iron in a water/methanol solution. *Environ. Sci. Technol.* 38(19): 5208–16.

Lyon, D., L. K. Adams, J. C. Faulkner, and P. J. J. Alvarez. 2006. Antibacterial activity of fullerene water suspensions: Effects of preparation method and particle size. *Environ. Sci. Technol.* 40(14): 4360–66.

Lyon, D. Y., J. D. Fortner, C. M. Sayes, V. L. Colvin, and J. B. Hughes. 2005. Bacterial cell association and antimicrobial activity of a C_{60} water suspension. *Environ. Toxicol. Chem.* 24(11): 2757–62.

Maynard, A., and R. Aitken. 2007. Assessing exposure to airborne nanomaterials: Current abilities and future requirements. *Nanotoxicology* 1:26–41.

Oberdörster, E., P. McClellan-Green, and M. Haasch. 2006a. Ecotoxicology of engineered nanoparticles. In *Nanotechnologies for the life sciences.* Vol. 5, *Nanomaterials: Toxicity, health and environmental issues,* ed. C. S. S. R. Kumar, chap. 2. Weinheim, Germany: Wiley VCH.

Oberdörster, E., S. Zhu, T. M. Blickley, and P. McClellan-Green. 2006b. Ecotoxicology of carbon-based engineered nanoparticles: Effects of fullerene (C_{60}) on aquatic organisms. *Carbon* 44:1112–20.

Phenrat, T., N. Saleh, K. Sirk, R. D. Tilton, and G. V. Lowry. 2007. Aggregation and sedimentation of aqueous nanoscale zerovalent iron dispersions. *Environ. Sci. Technol.* 41(1): 284–90.

Smith, C. J., B. J. Shaw, and R. D. Handy. 2007. Toxicity of single walled carbon nanotubes to rainbow trout (*Oncorhynchus mykiss*): Respiratory toxicity, organ pathologies, and other physiological effects. *Aquat. Toxicol.* 82(2): 94–109.

Tinkle, S. S., J. M. Antonini, B. A. Rich, J. R. Roberts, R. Salmen, K. DePree, and E. J. Adkins. 2003. Skin as a route of exposure and sensitization in chronic beryllium disease. *Environ. Health Perspect.* 111:1202–8.

Tong, Z., M. Bischoff, L. Nies, B. Applegate, and R. F. Turco. 2007. Impact of fullerene (C_{60}) on a soil microbial community. *Environ. Sci. Technol.* 41:2985–91.

Xia, T., M. Kovochich, J. Brant, M. Hotze, J. Sempf, T. Oberley, C. Sioutas, J. I. Yeh, M. R. Wiesner, and A. E. Nel. 2006. Comparison of the abilities of ambient and manufactured nanoparticles to induce cellular toxicity according to an oxidative stress paradigm. *Nano Lett.* 6(8): 1794–807.

6

Nano LCRA: An Adaptive Screening-Level Life Cycle Risk-Assessment Framework for Nanotechnology

Jo Anne Shatkin

CONTENTS

Integration of risk assessment with the study of life cycle stages of nano-technology-enabled products (NEPs) permits identification of critical risk assessment data needs.

National Nanotechnology Initiative, 2011

This chapter brings together many of the central concepts developed in the preceding chapters to arrive at the core theme of this book: adopting life cycle thinking and risk assessment for managing the hazards of nanomaterials and nanotechnology. Building on the basic understanding of risk assessment given in Chapter 2, it describes the concepts of adaptive management and screening-level analysis, and explains Nano LCRA (life cycle risk assessment), an adaptive screening-level risk assessment framework that incorporates life cycle thinking.

Nano LCRA is based on adaptive management and involves revisiting the framework and reevaluating prior decisions with new information. The screening-level, or qualitative, assessment process is targeted to today's decision making, that is, in the current state of uncertainty about the environmental and biological behavior of nanomaterials. This framework provides a sound scientific basis for sustainable nanotechnology development.

Following the discussions in Chapters 4 and 5 of the state of the science, it is clear that the scientific understanding of nanomaterial behavior in terms of health and environment remains at an early stage. It is not in its infancy, because much of what has been learned about the behavior of other types of particles and chemical substances is applicable to nanomaterials, but it is not matured either, due to the complexities of small-particle dynamics with which health and environmental scientists have little experience. The difficulties in measuring nanomaterials in relation to biological effects limits the necessary testing. This is in contrast to the rapid technology development that demands action now.

As we have discussed, nanotechnology applications are developing independently of a regulatory structure. Currently, regulatory decisions are being made on a case-by-case basis, a source of uncertainty that limits commercialization. The uncertainty in scientific assessment holds implications for managing the potential risks, since there are missing data for establishing thresholds, particularly for environmental exposures. The process of regulation may take years, and in its absence, agencies are developing guidelines and policies to protect workers and public health. However, nanotechnology risks need to be managed now, despite the lack of standardized methods for measuring exposure and risk. Hence, an adaptive approach—developed with the best available information and targeted to limited exposure and risk while learning more and improving management of uncertain risks—is the best path forward to protect health and the environment and not stifle innovation while the science is sorted out and the regulatory process is fought over.

The level of knowledge about nanoscale material impacts might be compared to the experience of learning to ride a bicycle. In this comparison, the current state of knowledge might require training wheels. It is probably beyond a tricycle, but certainly not confidently rolling down a road. To relate this level of understanding to risk assessment is somewhat like the understanding one gains from a screening-level or back-of-the-envelope qualitative analysis. Full quantitative risk assessment is comparable to a cyclist clipping her feet into clipless pedals and racing with the *peleton* in the Tour de France (minus the transfusions) or at least to a club ride at sunrise. It is not that risk assessment is so complex; it is that the level of effort, knowledge, and experience required to conduct full risk assessments is much more extensive, and for most nanomaterials, these data are not yet at the necessary level of quality and reliability. In other words, it makes sense to learn how to ride a bike before attempting to race with the best of the best. An adaptive approach, an iterative analysis with increasing levels of understanding, skill, and quantitation, presents a path forward for evaluating and managing the risks of nanomaterials. This path allows adaptation to new information, decision making under uncertainty, and a manageable process for identifying and prioritizing concerns about health and environmental risks.

Identifying the potential implications in technology development and evaluating their likelihood requires both a risk-based approach and life cycle thinking. The adaptive life cycle risk assessment (LCRA) approach is founded in the use of science for environmental decision making. It is also inherently a "win-win" economically, environmentally, and socially, the so-called triple bottom line (Elkington 1998). The approach is also a proactive way to manage technology.

An adaptive approach is not an alternative to regulation and standard setting; it is a path forward that collects the best available information early in the design process to ensure that new products, technologies, and materials are designed to be safe and sustainable. The approach described here is one that leads to safer product design in advance of regulation.

6.1 Adaptive Management for Nanomaterials Using Risk Analysis

The Nano LCRA framework applies *adaptive management* for nanomaterials; it adopts the tools of risk analysis and life cycle thinking to characterize the potential for exposure and risk to nanomaterials in specific applications. Adaptive management in this context means making the best decisions with the available information, ensuring that these decisions can be updated when new information becomes available, and ensuring that timely re-evaluations occur. Adaptive management *requires* building re-evaluation into the process. That is, one conducts an analysis initially when available information may be scarce, with the intention that the results of this initial analysis will involve conservative and protective actions to manage any identified risks. This drives the need for further data gathering and analysis to better characterize the potential for human and environmental exposure and risk.

Adaptive environmental management integrates environmental, economic, and social aspects of complex issues as an alternative to traditional, reactive management solutions. It represents an approach to managing complex systems with several key attributes relevant for nanoscale materials and nanotechnology. Designed for situations that are poorly understood and somewhat unpredictable, adaptive management identifies key uncertainties and conducts experiments to better understand and manage them. The main objective of adaptive management is to adapt and learn, improving the process or intervention. Responses are developed that represent opportunities to learn about uncertainty, and these responses are routinely revisited with the learning that has taken place. Conceived of in the 1970s as an approach for ecosystem management in forests and other complex systems, adaptive management is widely used in a range of situations with social as well as environmental complexity.

The concept of experimenting to learn about how to manage risk might seem disconcerting, but in fact this approach is fairly common. A problem is identified and evaluated, and a solution proposed, but often it is later learned that new problems arise from the solution, which in turn must be managed, and the initial solution also requires revision. Some argue that the world is so complex that introducing new technology is akin to conducting experiments that no one can fully grasp and manage (Giddens 1998). As a result, to achieve a sense of control over the environment, many of us are preoccupied with calculating and managing risks (Beck 1998), which inherently is adaptive management.

The current situation with nanomaterials and nanotechnology is uncertain and complex, begging for solutions that use the best available knowledge to address risks while continuing to focus on learning about the key variables affecting exposure, toxicity, and risk. The ecological and societal impacts of technology result from diverse uses in consumer products that are used by many people, and ultimately disposed of in the environment. Dispersed not only in spatial terms, technology also evolves so quickly that its rapid obsolescence means that more and more electronics enter the environment, and the technological impacts require nearly continual adaptation. Managing the human and environmental aspects of technology in an adaptive manner means that society as a whole can benefit from learning how best to manage technological risk, but it also empowers those who are preoccupied with understanding and managing risks to participate.

There is one aspect of adaptive management that is not well suited to nanotechnology, that is, the spatial scale. In a global economy, with technology and nanomaterials being introduced across economic sectors and geographic boundaries, it is difficult to envision a global scale for adaptive management. Most effective environmental management occurs at a local or regional scale, where the key stakeholders and participants are committed to a solution because they are directly impacted by it. Adaptive-management approaches for nanotechnologies might not be universal, and they might be implemented differently for different sectors, allowing additional learning from the diverse implementation. Perhaps by the time most nanotechnologies achieve global integration, risk assessments for them will be fairly routine.

6.2 Screening-Level Risk Assessment and Adaptive Management

The philosophy for an adaptive framework incorporating life cycle thinking into risk assessment for nanotechnology is based on the view that an early *screening-level analysis* with considerable uncertainty and a lack of available data will have two valuable outcomes: It develops early information needed to make sound decisions about product design and development, and it also offers an opportunity to make decisions amid uncertainty. There are three unique components to the Nano LCRA framework: *adaptive management, life cycle thinking*, and *screening-level risk assessment*. The framework is applied as a screening tool to identify and prioritize potential risk concerns, and to develop strategies for investigating and managing them further.

Before presenting a step-by-step explanation of the Nano LCRA framework, first we consider the rationale, main principles, and how such an approach came to be developed, including some of the significant issues it addresses for risk assessment of nanomaterials. While detailed risk analysis and life cycle impact assessments require a lot of data, the needed data for emerging technologies are generally not available. Some argue that complete data sets are necessary before nanomaterials can be comfortably allowed into commercial products that will enter into the environment. However, it has been demonstrated that combining available information with professional judgment can be used to guide decision making and to prioritize the gathering of more complete information, sometimes called "back of the envelope."

An excellent reference is the book, *Consider a Spherical Cow: A Course in Environmental Problem Solving* (Harte 1988). Harte describes how solutions to complex environmental problems can be simplified by initial estimation. As suggested in the title, defining the surface area of a cow can be estimated by assuming it is spherical (Harte 1988). This approach is particularly valid under conditions of uncertainty. Conducting screening-level analysis does not suggest that the available information is adequate to answer all questions, but it does allow estimation to approximate the significance of potential impacts. Detailed quantitative estimates may not be calculable initially, but order-of-magnitude estimates may be, and this is often adequate for informing interim control strategies or product design decisions. Stepping through the analysis identifies what is unknown, and important to know, in terms of: who may be exposed and how; how significant those exposures may be; and where in the environment concerns may exist regarding the presence of nanomaterials. In particular, it is clear from looking at a few examples that there may be limited or no exposure to nanoparticles or nanomaterials at certain stages across their life cycles. For example, manufacturing nanotubes in an enclosed process limits exposure to them. A screening-level analysis can help to document and identify which life cycle stages may be of concern and require further investigation and management policies to prevent exposure. The screening-level approach of Nano LCRA adopting life cycle thinking into risk analysis helps to establish priorities for future work.

One key issue to address is how a screening-level approach can indeed satisfy the concerns of those who wish to adopt a precautionary approach, i.e., conducting a full analysis of all aspects of a substance before it is developed into products. As Tukker (2002) points out, there is an inherent difference in philosophy regarding sustainability and the level of confidence in technology management, i.e., between those who prefer an analytic versus a precautionary approach. Adopting the Nano LCRA approach for risk analysis can address the concerns of these divergent views. The reason is that the initial screening provides some level of confidence because, in situations of greater uncertainty, risk management and decision making will address the uncertainty by adopting more risk-averse measures until better information is developed. This approach motivates further actions to better characterize and understand health and environmental impacts. Further, Nano LCRA takes a broader approach to managing risk than traditional hazard assessment, so overall it is more protective of health and the environment.

To summarize, adopting a screening approach guides early-stage sustainable technology development when there is not enough information to make definitive determinations of the health and environmental risk associated with new materials and technologies, yet provides a clear path forward to incorporate the available information into risk management and decision making. The available data is supplemented by adopting tools from *risk analysis* for addressing *uncertainty*. The *adaptive-management* aspect allows learning from the analysis and subsequent decisions to be made that make sense today, allowing for the possibility that these decisions may need to be adapted as new information becomes available. *Life cycle thinking* means that unintentional releases and the potential for indirect exposures are broadly considered with reference to what is known from past examples. *Risk assessment* means formal consideration of both the likelihood and the significance of potential exposure.

Even without a quantitative evaluation, a lot can be learned about potential risk to inform nanotechnology management going forward. Even when quantitative estimates of risk are made, they are just that—estimates—and they tend to be very conservative. By making assumptions to quantify risk, consistent procedures for evaluation allow characterization of new materials similar to substances with greater familiarity. Early consideration of the potential for exposure and risk will guide science to better understand the potential risks associated with nanomaterials.

6.3 Nano LCRA: An Adaptive Screening-Level Life Cycle Risk-Assessment Framework for Nanotechnology

Figure 6.1 shows the proposed Nano LCRA framework for nanotechnologies. The concepts are quite similar to Comprehensive Environmental Assessment and the other assessment approaches described in Chapter 7 that conceptualize the life cycle of a product when conducting exposure assessment.

The key difference is that the adaptive risk framework is a screening tool to identify and prioritize key health and environmental issues that may be applied at a very early stage of nanomaterial development when little information is available for risk assessment. The first iteration of the Nano LCRA framework identifies what information is really needed to make a better decision; however, it also allows early decisions to be made, with the intention that they will be revisited when more information becomes available. This dynamic approach is applicable to a broad array of hazards, materials, and technologies. In fact, there is nothing specific to

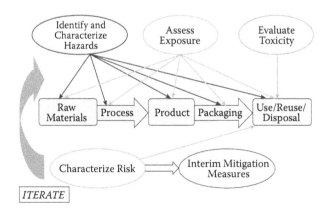

FIGURE 6.1
Proposed NANO LCRA framework for nanotechnologies.

the nanoscale about it. It allows an evaluation to occur at each stage of the supply chain; it can be equally applied to a raw material producer or to a downstream user of a product containing a nanomaterial in a composite, or both.

In a situation of significant data gaps, *exposure* is the place to start the evaluation. Of course, understanding the toxicity of new materials is important, but time and significant resources are required to develop toxicity data. Considering the potential for exposure first makes sense. It means detailed toxicological evaluations for materials or products where exposure may be low or nonexistent (for example, when carbon nanotubes are embedded in a solid composite material) and thus may not be required. Instead, resources are focused on toxicology evaluations in the life cycle stages when potentially significant exposure to nanomaterials may occur. Without exposure evaluation, however, these concerns will not even be identified. The analyses are systematic evaluations to ensure comparability of one type of material, or exposure, to another. The comparisons across materials, products, and technologies are the initial steps of quantifying exposure and risk.

Moving on now into the specifics of each stage of the process will help clarify the overall significance of adopting this adaptive approach. The idea is to step through the risk-analysis process, and for each stage of the life cycle to conduct a mini hazard assessment and an exposure assessment—the first two steps of risk analysis. These steps identify where to focus future activities in order to evaluate the toxicology and to conduct a risk characterization. The early evaluation can help to identify mitigation measures that ought to be used, but it also will pinpoint what information is missing in order to develop a more detailed characterization of risk. That information then becomes part of an adaptive approach.

6.3.1 The Ten-Step Nano LCRA Framework

1. Describe the life cycle of the product.
2. Identify the materials and assess potential hazards in each life cycle stage.
3. Conduct an exposure assessment for each life cycle stage.
4. Identify stages of life cycle when exposure may occur.
5. Evaluate potential human and nonhuman toxicity at these key life cycle stages.
6. Analyze risk potential for the selected life cycle stages.
7. Identify key uncertainties and data gaps and communicate findings.
8. Develop mitigation/risk-management strategies and next steps.

9. Gather additional information.

10. Iterate process, revisit assumptions, and adjust evaluation and management steps.

The Nano LCRA framework starts at the beginning of the manufacturing process. If you are a raw material producer of nanomaterials, then the starting material may not be a nanomaterial. For example, if you manufacture carbon nanotubes, you may start with iron and carbon black as raw materials. Step 1 describes in detail the manufacturing process and life cycle of the product, from generation to ultimate disposal. The framework is flexible enough to evaluate products at any stage of the supply chain. The product may be: a nanomaterial; the incorporation of a nanomaterial into a composite; the incorporation of a composite material into a product; a component of another product, such as a switch; and/or the final assembled unit, such as an airplane.

In step 2, the hazards associated with each stage of the product life cycle from premanufacturing to packaging the product are identified and characterized. The analysis proceeds to how the product is used, and how those uses might lead to new hazards when the product or material reaches its end of life, including how it is disposed of or recycled. The potential hazards (e.g., contamination, reuse) associated with all of these steps are identified using all available data about material composition and characterization. Hazards might include: the presence of free nanoparticles, fire, or explosion hazards, equipment failure, and accidental or intentional releases of nanomaterials. In particular, this step identifies those nano-specific hazards at each life cycle stage.

Step 3 conducts an exposure assessment looking entirely across the life cycle to identify potential scenarios. First, exposures to raw materials are assessed: For example, is mining involved to obtain raw materials? How are the raw materials handled and managed once they reach a processing facility? Who might be exposed, how frequently, and to how much? Is there no exposure to nanomaterials in the raw materials?

Next we look at the process of manufacturing a product: When and what types of exposure might occur? Can free particles be released into the air? Is the process entirely enclosed, so that no exposure to particles could occur unless a problem arises? Is a composite material cut or heated? Does the process involve touching the material or otherwise coming into contact with it? Is a lot of waste material produced that might be put in the trash or otherwise disposed of outside of the facility? Does the process allow particles to be entrained into a ventilation system and released elsewhere?

The framework continues to step through the exposure scenarios that could occur once a product is produced. Next is to focus on the packaging step in manufacturing and evaluate whether exposures might occur. If the product is a powder, the packaging step might be the most significant. In some manufacturing designs for the packaging step, the only opportunity for exposure may be if the equipment requires periodic maintenance or breaks and needs repairs.

With the first three steps completed—the life cycle, hazard assessment, and exposure analysis—step 4 evaluates which parts of the life cycle have the greatest potential for exposure to nanomaterials and should be the focus of the product risk assessment. This requires careful consideration of all potential pathways, the form of the material at that stage, and the likelihood that human or environmental exposure could result.

Step 5 then identifies the available toxicology data to assess the types of potential adverse effects. As discussed in Chapter 4, the information currently available about the levels of nanomaterials that may adversely affect health is complex and not easy to interpret. The screening-level evaluation may identify some preliminary concerns, but it may also include data gaps, particularly for specific routes of exposure, such as oral ingestion of nanoparticles.

Step 6 compares the available toxicology data with the exposure and hazard data to characterize risk. Because of data gaps, it is likely that early iterations will be more qualitative than quantitative in describing risk. One way to get a sense of the potential for significance of risk is to compare it to other risks, including alternative technologies and materials. Another is to obtain expert input regarding the potential risk. Another tool is to use bounding analysis, i.e., using estimates or probabilistic tools to generate minimum, most likely, and reasonable high-end estimates (these are more relevant than worst-case scenarios) of risk levels.

Step 7 evaluates the level of confidence in the assessment by identifying key sources of uncertainty and then documenting the level of confidence in the results. Whether this is a quantitative exercise depends on the availability of data. With little data, it sometimes makes more sense to use "low-medium-high" as a scale describing the confidence of the assessment, the data itself, and the general weight of the evidence. When more detailed information becomes available, the uncertainty/confidence assessment becomes increasingly quantitative.

Step 8 applies what is learned from the analysis to develop alternatives for how to manage the risks. Using information from the analysis, control measures may be developed. It is important to realize that as an adaptive management process, these may be interim steps until more detailed information is developed. However, the management strategy is informed by the prior steps of the assessment. The analysis has likely identified some uncertain and crucial areas for further investigation. The management strategy includes a plan for addressing the uncertainties, and this becomes a living document, i.e., it is updated as the framework is reiterated.

With management/mitigation measures implemented, the last two steps repeat the entire process. This may include gathering additional data, but it certainly includes evaluating the efficiency of management efforts to mitigate risk, and it also identifies the next set of priorities to be addressed. Some data gathering may take time, so appropriate intervals are set for iterating the LCRA and updating the mitigation measures in the management strategy. Of course, each of these steps has an outreach and communication component.

Thus, conducting a screening-level risk evaluation has helped to focus on the key exposure and risk issues of that particular material or technology and its life cycle. The process has identified the next steps, in particular identifying additional data needs to more accurately quantify risks at specific life cycle stages, gathering more information, or implementing mitigation measures: for example, changing the packaging process to prevent exposure, or perhaps conducting toxicity evaluation for exposures at a particular life cycle step. Once the additional information is developed, the steps of the Nano LCRA framework are revisited to evaluate the impact of the mitigation measures, which could indicate, for example, that the next greatest concern regards the raw materials and identifies the next set of priorities.

The output of applying the framework initially may be more qualitative than quantitative in describing the potential risks. It will describe the identified stages of exposure to nanomaterials that might occur for the entire life cycle of a nanomaterial or nanotechnology. Using this qualitative approach creates a sound basis for decision making as well as a set of data needs that require greater focus going forward to characterize the risk. Importantly, the early application of the framework prioritizes data collection to inform the most significant data gaps from a risk perspective.

For example, a nanopowder has greater potential for inhalation exposure, dermal contact (by touching the powder), and ingestion exposure (by a worker or consumer accidentally putting contaminated hands in or near the mouth). A toxicology evaluation can then focus on those pathways and on the direct exposure to the powder. If toxicity information needed for evaluation is not available, the analysis has still identified an important next step for characterizing risk: that a more detailed evaluation of this particular exposure scenario is needed. That in itself is a significant finding: enough initial information to allow "thinking through" how to mitigate, or even prevent, exposure during that step. This is particularly useful during product design and development. Thus, even in the absence of key information, the process still yields the ability to characterize risk and document the characterization for management options.

One of the benefits of conducting a screening analysis using the Nano LCRA framework is that it identifies the important information to document what is known, but also to learn what is not known and may require further investigation in order to make sound science-informed decisions. Stepping through this process may lead to the discovery that there is a potential for exposures that are unacceptable, even if these cannot be quantified. This may lead to changes in the manufacturing process or redesign, even before any equipment is purchased. As discussed in Chapter 3, this is an important finding and an important benefit (both environmental and economic) of using this framework.

The Nano LCRA framework has direct application to health and safety concerns. Applying this process of looking at the life cycle of a material considers health broadly, including occupational exposures, consumer exposures, and the health of the environment. Considering exposure first will help to identify hazards and assess exposures in a way that will elucidate safety concerns.

Another feature is that the framework leads to management decisions linked directly to the potential for risk. The rationale for taking certain actions over others is clearly identified. Whether this becomes the basis for a regulatory policy, a company policy, or a voluntary activity that can be used to certify technologies, this process demonstrates proactive management of risks, both within an organization and to stakeholders. It incorporates clear communication of efforts taken to mitigate risk and the rationale for doing so. Further, because risks are assessed in real time, for novel materials early in their innovation process, adaptive approaches allow early decisions for managing risks based on sound science, even under conditions of uncertainty.

The Nano LCRA framework steps sequentially through processes across the product life cycle, evaluating risk at each step. It focuses on exposure potential and addresses the "worst things first" mentioned at the beginning of the book. It allows prioritizing concerns. As an adaptive approach, it iterates new information with changes to a process or product. It allows great flexibility in designing management systems to address potential exposure.

The Nano LCRA framework is transparent because it documents the assumptions and analysis and allows comparison of different products and processes amid uncertainty. During the research and development process, there is still an opportunity for testing alternative manufacturing methods that could be used when scaling up from research and development to manufacturing. Further, using the framework allows comparison of the types of environmental health and safety issues that would need to be addressed under alternative manufacturing scenarios: for example, evaluations to assess whether to capture nanomaterials in a liquid matrix or in an air filter, what types of management would be required for those materials in one media versus another, or whether a process can be designed to capture the materials for reuse versus disposal. Finally, the Nano LCRA framework offers a simple, proactive, dynamic approach that can be reevaluated regularly.

6.3.2 Examples Using the Nano LCRA Framework

Some products are intended to create nanoparticle exposure during use: sunscreen, for example. The active ingredient in sunscreen may be a nanoparticle such as titanium dioxide that blocks ultraviolet rays from reaching the skin; thus, exposure is intended during use in this application.

In assessing product use, it is important to consider what happens to the materials in a product after its disposal, the post-consumer use. With drugs, for example, the American Medical Association (AMA) used to recommend that when the course of treatment was finished, the unused prescription drugs should be flushed down the toilet to limit the potential for accidental exposures by others and thus prevent their consumption by people for whom they had not been prescribed. However, after pharmaceuticals are flushed down the toilet or the drain, they can then flow into a wastewater treatment system that is not necessarily designed to capture them. Thus, after the water is treated, the drugs can be released into other water bodies. With a number of pharmaceuticals beginning to appear in water bodies, the U.S.EPA has recently released guidance recommending that people dispose of their used prescriptions in the trash. A recent survey by the U.S. Geologic Survey measured a number of pharmaceuticals in streams and other water bodies downstream of wastewater treatment systems. It also measured substances such as caffeine that were clearly associated with human use (Kolpin et al. 2002). Thus, if nanomaterials are part of drugs, they too could be released into waters. Even if unused portions of prescription drugs are thrown into the trash, they can still enter the water cycle when excreted in the urine or feces of the people using them.

Materials such as product packaging that are not necessarily recycled also go into the trash, and their ingredients can also make their way into the environment. Sometimes trash is incinerated, and materials can be released into the air or become associated with the ash after burning. Trash that goes to a solid waste landfill can break down and release components into the environment via gases or liquids (leachate). Modern landfills treat the leachate, but it is currently unclear whether nanoparticles would be adequately captured during that treatment.

There are many other pathways for potential exposure. As discussed in Chapter 3, cerium oxide added to fuel could be released with engine exhaust. Putting silver nanoparticle coatings on fabrics in public areas will eventually release the silver to the environment during laundering activities, as fabrics break down or are discarded, or during reupholstering of furniture. Nanomaterials added to building materials for self-cleaning or antibacterial use, electrical conductivity, or for strengthening structures will most likely eventually make their way into the environment.

Many materials that we use now, particularly for food packaging, are recycled, and if these materials contain nanoparticles, what will be the fate of the nanoparticles when the material is recycled? What types of exposure could occur during recycling and in these materials' secondary uses? It is important to determine how exposure to nanoparticles might occur during the use, reuse, and disposal steps.

This has all been fairly abstract thus far, so let us walk through a couple of examples.

6.4 Quantum-Dot Additive to a Coating Case Study

- Step 1: Describe the Life Cycle

  QD Vision, a nanomaterial development company that makes quantum dot (QD) lighting solutions, has an idea for making a specialized coating product that has unique optical properties. The life cycle of the QD-based coating may be described in stages; these begin with three manufacturing stages and four postmanufacturing stages. First is the synthesis of the quantum dot. The reactants are added to a vessel in a vented hood, where the QD cores are synthesized as colloidal semiconductor particles in elevated-temperature organic solvents. The cores are then overcoated with another semiconducting layer, selected for its band-gap energy relative to the core. Next, the dots are encapsulated within a polymer to protect the optical properties. The encapsulated dots are introduced to the coating formulation, stored as a liquid in 500-ml safety-coated glass containers, and shipped to the customer.

  The next stage is the application of the coating to an object. Various forms of application may be used, including brushing and spray application. The coated material is used, and once it is no longer in use, it is disposed of, either in a landfill or incinerator.

- Steps 2 and 3: Hazard Identification and Exposure Assessment

  The potential hazards associated with QDs used in a coating product will vary at each stage. Production occurs in a well-designed and well-controlled manufacturing environment. There are health and safety hazards associated with the precursor materials, including the metallic components of the raw core and shell, the formulated core and shell components in organic solvents, the glass reaction vessels when working with them at high temperature, and the physical safety risks associated with manually transporting the reaction vessels from the reaction hood to the cleanup hood and back. The impacts of encapsulation on optical performance of the coating was an outstanding question in the phase of product development during our study, so we investigated the potential impacts on exposure and health risks to both unencapsulated and encapsulated dots. More handling is required to encapsulate the dots, but the resultant product is three orders of magnitude larger in diameter, potentially limiting the probability of nanoparticle exposure in latter stages. The question of whether encapsulation reduced release of QDs from the polymer coating was also investigated. The formulation step has the associated potential risks of working with QDs in solvents as well as with monomer (both nano and non-nano) substances. Formulations contain solvents and quantum dots on the order of about 10^{18} dots in a batch.

Next, the packaged coating is shipped as a liquid. Because of the presence of flammable solvents, the sample is shipped as a hazardous material and packaged according to Department of Transportation/International Air Transport Association (DOT/IATA) regulations for such materials. The spray coating potentially releases QDs, polymers, and solvents as an aerosol in the breathing zone of the applier. Spraying or brushing the liquid coating creates a short-term potential for exposure to the nano and non-nano components through direct contact with skin or clothing during application.

In the use of the coated product, QDs embedded in the substrate surface could potentially be released via direct skin contact with the user, who may not be aware of the presence of a nanomaterial. Environmental conditions and mechanical wear of the coating can potentially change the risk of exposure at this stage. Finally, the disposal of the liquid product could be to water, landfill, incineration, or other land disposal. The risk at the end of life is the potential for release of QDs into the environment, where any number of potential human or environmental receptors might contact the QDs or the polymer directly.

- Step 4: Identify the Life Cycle Stages for Risk Assessment

  Because the manufacturing steps were highly controlled in a fume hood and safety equipment was worn by personnel, the critical life cycle stages for the Nano LCRA were the spray application, weathering of the coating, and end of life. In the spray application, inhalation and dermal exposure by the user and passersby were the potential pathways to evaluate. For weathering, we analyze the situation where either nanoparticles, or particles attached to coating, or quantum dot metallic components could directly come into contact with a person or indirectly be carried off by touching or from weather-related disturbance. The ultimate disposal of the formed coating, or its liquid precursor, is the third exposure scenario, where improper disposal might contaminate water resources or other human or environmental pathways.

- Step 5: Toxicology

    There is an emerging body of literature about the potential toxicity of certain specific QD materials to human and environmental species. No information was located for the specific core/shell configuration, but the toxicity of the raw metallic ingredients is known, and therefore there is a base of knowledge to draw from. One purpose for the Nano LCRA was to determine which of the specific life cycle stages are useful for conducting toxicology experiments to inform subsequent study design.

- Step 6: Risk Characterization

    There was a very low likelihood of exposure from the manufacturing steps, since these had been designed to limit exposure. Accidental release is the only viable pathway. Encapsulating the dots reduced inhalation exposure risk and oxidation potential for the dots. The initial risk characterization found the need to assess the potential exposure to nanoparticles during spraying of the coating and following weathering in the natural environment. A series of experiments in a modified clean-room setting was unable to detect the presence of quantum dots during spraying and aggressive challenges to weathered coating (Shatkin and Coe-Sullivan 2012). Even with vigorous efforts to release them, none were detected, indicating very low exposure potential, below the limit of detection and background levels.

- Step 7: Identify the Uncertainties and Important Data Gaps

    Remaining data gaps include the end-of-life exposure scenarios. We developed a series of safe handling procedures, including a Material Safety Data Sheet, that specifically explain the need to responsibly dispose of the liquid material. Calculations suggested that the coated material at end of life would not create a health or environmental hazard. However, we did not test every scenario, and end-of-life impacts remain a source of uncertainty. The toxicology of the formulated dots, even in their encapsulated state, is a critical data gap. However, the low exposure levels reduced concerns about toxicology. Still, if the material becomes commercially available in large quantities, it may be necessary to quantitatively assess risks.

- Step 8: Develop the Risk Management Strategy

  The product manufacturer already had developed safe handling procedures for each of the manufacturing stages, based on other product lines. These were reinforced for this developing product and its applications. Safe handling procedures were developed for the downstream coating applicant that avoided direct contact, even though our product testing did not suggest the need for this layer of protection. Precautions for protecting the coated material were also part of the strategy. Finally, the safe handling procedures prohibited the disposal of the final product in an unsafe manner. However, this requires voluntary aspects on the part of the user, which the manufacturer was not in a position to control.

- Steps 9 and 10: Adaptive Aspects

  As methods improve and are standardized, some additional exposure testing may be warranted, as not all conditions could be represented. Further, it would be helpful to know whether the composite material has associated toxicity to health or the environment. Most studies suggest that some toxicity from quantum dots is due to the oxidation of shell/core material and the release of metallic components, but the data regarding toxicity of specific QD formulations is not quantitatively defined. Risk could be lowered by altering the formulation, but the risk for the final formulation was already low due to the lack of exposure, even during environmental uses.

6.5 Nano Iron Case Study

- Step 1: Describe the Life Cycle

  A start-up company is located in a shopping mall with office space, two small labs, warehouse space outfitted with an air-handling system, several benches for mixing materials, and several types of high-energy machines used for generating and processing nanoparticles. The main process uses a proprietary enclosed plasma arc process to generate nanoparticles of various metals. The product in fabrication is nanoscale iron particles. These are made from iron pentacarbonyl and argon gas. Iron pentacarbonyl is heated to become gaseous. The argon gas carries the iron pentacarbonyl to the high-energy plasma, which cleaves the iron-carbon bonds and produces iron nanoparticles. The argon gas carries the iron nanoparticles to a collection chamber, which is surrounded and cooled by liquid nitrogen. The iron is reactive in air; that is, it very quickly forms iron oxide (rust), and the speed of this reaction can cause the iron to spontaneously combust, so argon gas continues to flow over it as it cools to keep oxygen out of the collector. The iron powder is kept under gas and transferred to a machine that condenses the powder to remove the space between the particles and prevent them from aggregating into larger particles or from forming iron oxide. The surface of the condensed powders forms a 25-nm thick layer of iron oxide. The pressed powder is used for treating groundwater contaminated with chlorinated solvents.

The powder is immersed in high purity water diffused with nitrogen, packaged in an airtight container, and shipped to the field under nitrogen gas. Upon arrival, the material is mixed with a dispersing agent and injected into the deep groundwater through a series of airtight well encasements. The iron begins to travel with the groundwater and quickly breaks down the tetrachloroethylene and trichloroethylene (PCE and TCE) until it locates enough oxygen molecules to form iron oxide, which is poorly soluble and adheres to soil particles.

- Steps 2 and 3: Hazard Identification and Exposure Assessment

  At the production stage, iron pentacarbonyl is flammable and is used in a high-energy environment. The production of iron nanoparticles creates a combustion hazard, and if the iron particles spontaneously combust, they could break the glass container at the end of the column, releasing nanoparticles into the work area. The removal of the powder at the end of the process creates a hazard because it must be handled and could spontaneously combust. The packaging step requires manual handling of the powder on a bench, and so could include dermal contact with the powder, which could lead to inhalation or ingestion of nanoparticles if personal protective equipment is not worn properly. Although immersed, the iron nanoparticles could be released during transfer to the field. Field workers using the iron could come in contact with the iron particles. The iron could travel far from the site underground once injected into groundwater.

- Step 4: Identify the Life Cycle Stages for Risk Assessment

    This step finds that the highest potential for exposures is in the raw material stage, for iron pentacarbonyl, and in the post-production packaging phase, where significant handling of the iron nanopowder can lead to exposure. Accidental release from manufacturing process upsets, during shipment, and in the field are also exposure pathways of potential concern. The potential for environmental exposure when the iron is released into groundwater requires evaluation.

- Step 5: Toxicology

    Iron pentacarbonyl is highly toxic when inhaled, and can be absorbed through the skin. Information regarding effects associated with exposure to iron nanoparticles can be gleaned from studies of poorly soluble dusts, but there are few studies for iron dust only. As described in Chapter 5, zerovalent iron particles are associated with the generation of reactive oxygen species, a common toxicity mechanism for nanoparticles, and toxicity to bacteria and fish. *In vitro* studies report cellular toxicity. Potential secondary effects from the iron particles in groundwater could go undetected, but may affect downstream species such as fish.

- Step 6: Risk Characterization

    The synthesis process has both physical and toxicological hazards associated with the handling of iron pentacarbonyl and iron nanoparticles. The packaging step has potential for exposure to unbound nanoparticles, which may spontaneously combust. Field workers could be exposed to nanoparticles in preparing for deep well injection. There is potential for downstream environmental exposure from use of the iron nanoparticles.

- Step 7: Identify the Uncertainties and Important Data Gaps

    Exposure levels in the packaging process are not measured, and health effects associated with those exposures are poorly characterized. The frequency and intensity of process upsets and accidental releases is a data gap to address. The behavior of the iron nanoparticles in groundwater remains a key area of uncertainty. These data gaps can be addressed by adopting measures to mitigate the potential exposure or by initiating studies to better characterize their potential significance.

- Step 8: Develop the Risk Management Strategy

  Many steps must be taken to improve the safety of the manufacturing process, including encasing the equipment to limit injury potential; working with iron pentacarbonyl in a ventilated environment, such as a fume hood; and building protective layers around the gas handling in and out of the plasma generator to avoid direct contact with the nanopowder. The packaging step needs to be part of the same enclosure to prevent the possibility of spontaneous combustion and human exposure to the nanoparticles. Risks from field use can be mitigated through careful communication and worker training. Information on the behavior of iron nanoparticles in groundwater was not located.

- Steps 9 and 10: Adaptive Aspects

  Several unknowns require more detailed evaluation. Areas for further research include: inhalation, dermal exposure, and ecological toxicity of iron nanoparticles; particle concentration in workplace air; and the environmental behavior of nanoscale iron in groundwater, particularly in the presence of solvents. Workplace enclosures should be installed and the process revisited. Finally, as changes are implemented in step 8, the process is reiterated. Steps 9 and 10 require gathering additional information, iteration of the prior steps, and adjustment of the evaluation and management steps.

6.6 Summary

The Nano LCRA framework is an assessment and management tool that is particularly suited to early-stage sustainable product development. The Nano LCRA framework is iterative and adaptive, allowing decision making under uncertainty, and presents a path forward to address the uncertainties. It is a framework that applies now, to our current level of understanding for most nanomaterials, and can be broadly applied to any emerging substance. The novel aspects of nanomaterials require adaptive management—we must make decisions today, but today's decisions may not be in line with our thinking tomorrow. There is substantial overlap between the Nano LCRA framework proposed here and the frameworks discussed in Chapter 7, which are also focused on understanding the occupational and environmental risks from nanomaterials.

References

Beck, U. 1998. The politics of risk society. In *The Politics of Risk Society*, ed. J. Franklin, Chap. 1. London: Polity.

Elkington, J. 1998. *Cannibals With Forks: The Triple Bottom Line of 21st Century Business*. British Columbia, Canada: New Society Publishers.

Giddens, A. 1998. Risk society: The context of British politics. In *The Politics of Risk Society*, ed. J. Franklin, Chap. 2. London: Polity.

Harte, J. 1988. *Consider a spherical cow: A course in environmental problem solving*. Sausalito, CA: University Science Books.

Kolpin, D. W., E. T. Furlong, M. T. Meyer, E. M. Thurman, S. D. Zaugg, L. B. Barber, and H. T. Buxton. 2002. Pharmaceuticals, hormones, and other organic wastewater contaminants in U.S. streams, 1999–2000: A national reconnaissance. *Environ. Sci. Technol.* 36(6): 1202–11.

Shatkin, J., and S. Coe-Sullivan. 2012. Streamlined Life Cycle Risk Assessment to Evaluate Human Health and Environmental Risks of Novel Nano-enabled Products. *In press.*

Tukker, A. 2002. Risk analysis, life cycle assessment: The common challenge of dealing with the precautionary frame (based on the toxicity controversy in Sweden and the Netherlands). *Risk Analysis* 22(5): 821–32.

Further Reading

Adams, J. 1999. Review of *The Politics of Risk Society*, ed. J. Franklin. *J. Forensic Psych.* 10(1): 203.

Cordis. 2004. Towards a European strategy for nanotechnology. *COM* 2004: 338. http://cordis.europa.eu/nanotechnology/actionplan.htm

Davis, J. M., and W. H. Farland. 2001. The paradoxes of MTBE. *Toxicological Sciences* 61:211–17.

Oberdörster, E., P. McClellan-Green, and M. Haasch. (2006). Ecotoxicology of engineered nanoparticles. In *Nanotechnologies for the life sciences*. Vol. 5, *Nanomaterials: Toxicity, Health and Environmental Issues*, ed. C. S. S. R. Kumar, chap. 2. Weinheim, Germany: Wiley VCH.

Shatkin, J. A. 2005. Risk assessment: Informing the development of beneficial technology. Presented at the Advancing Beneficial Nanotechnology Foresight Conference, San Francisco. http://www.foresight.org/conference2005/presentations/shatkin.pdf

Sweet, L., and B. Strohm. 2006. Nanotechnology: Life-cycle risk management. *Human and Ecological Risk Assessment* 12(3): 528–51.

7

Comprehensive Environmental Assessment and Other Life Cycle-Based Approaches for Assessing Nanotechnology

J. Michael Davis

CONTENTS

The idea of integrating life cycle thinking and risk analysis has been around for several years, and in recent years this idea has been proposed specifically for evaluating nanotechnology risks (Cordis 2004; Shatkin 2005; Sweet and Strohm 2006; Davis 2007; U.S.EPA 2007; Sass 2007). However, implementation of such proposals has been limited due to the relatively immature state of the science regarding the implications of nanotechnology for human health, ecological receptors, and other types of impacts. Thus, it is not surprising that much of the work in this area has a hypothetical flavor and comes up short in arriving at conclusions about the risks of nanotechnology. Without adequate information, risk assessment efforts cannot get very far.

That said, it is still important to have an assessment approach in mind as a guide to filling the information gaps surrounding nanotechnology. The Nano LCRA (life cycle risk assessment) framework described in Chapter 6 illustrates one such assessment approach for nanomaterials; other assessment approaches that incorporate a life cycle perspective are discussed in this chapter, with a particular focus on one approach known as Comprehensive Environmental Assessment (CEA). Although all of these approaches include a life cycle orientation, they differ in some respects, as will be noted in the following discussion.

7.1 Comprehensive Environmental Assessment

The idea of Comprehensive Environmental Assessment (CEA) was first developed in reference to fuels and fuel additives (Davis and Thomas 2006), although its applicability to other technological issues, including nanotechnology, has been apparent (Davis 2007). Its origins in relation to fuels/fuel additives (F/FAs) owes a great deal to the Alternative Fuels Research Strategy (U.S.EPA 1992) that was developed by the U.S.EPA's Office of Research and Development to lay out a framework for assessing the benefits and risks of various F/FAs. In essence, both the Alternative Fuels Research Strategy and the CEA approach combine a life cycle perspective with the risk assessment paradigm.

The advantage of a life cycle perspective is that it allows a broader, more systematic examination of the trade-offs associated with a product. This point is well-illustrated by the case of methyl tertiary butyl ether (MTBE), a fuel additive that has been widely used to increase the oxygen content and octane number of gasoline. As discussed in Chapter 3, during the 1990s, MTBE use grew dramatically in the United States, mainly in response to provisions in the 1990 Clean Air Act Amendments that called for the use of oxygenates in gasoline to address certain air quality problems. Although MTBE was at one time used in approximately one-third of U.S. gasoline, its use declined precipitously because of concerns about its potential to contaminate water resources when leaking from underground fuel storage tanks (U.S.EPA 1998, 1999). Thus, a product that was intended to improve air quality ended up being a source of water pollution.

The Alternative Fuels Research Strategy (U.S.EPA 1992) presciently warned about potential problems with MTBE (and a related oxygenate, ethyl tertiary butyl ether [ETBE]) when it stated: "Compared to gasoline, the ethers MTBE and ETBE have relatively large aqueous solubilities and would likely leach more rapidly through soil and groundwater. Also, limited data suggest that ethers may be persistent in subsurface environments." And, "Very little is known about emissions and releases from MTBE and ETBE storage and distribution, making this area an appropriate target for research. Effects on existing equipment and control...need to be evaluated" (U.S.EPA 1992).

As it turned out, the propensity of MTBE in gasoline to leak from underground fuel storage tanks and thus foul groundwater proved to be the Achilles heel of this product. But correctly anticipating this problem was not a fluke or coincidence; rather, it was the result of a collective effort by U.S.EPA scientists to think through various implications of MTBE and other F/FAs in relation to the entire life cycle of the fuels, not just their intended end use. The CEA concept extends and formalizes the approach that was used in the Alternative Fuels Research Strategy.

7.1.1 Features of Comprehensive Environmental Assessment

The CEA approach consists of both a *framework* and a *process*. As will be explained in the following discussion, the framework provides a holistic, systematic way to organize information about complex environmental issues, while the process affords a transparent and structured means to engage diverse expert and stakeholder perspectives in judging the implications of such information. In addition, CEA can be understood as a *meta-assessment* approach that builds on existing assessment methods and, to the extent possible, the results of available analyses.

7.1.1.1 CEA as Framework

Figure 7.1 highlights key features of the CEA framework, beginning with the *product life cycle*. Although not routinely included in life cycle assessments (LCAs), the product life cycle generally starts in the research and development laboratory. Other stages of a product life cycle can include processing of feedstock materials along with the manufacturing, storage/distribution, use, and disposal/recycling of the product. At each of these steps, some potential may exist for releases/emissions of materials into the environment. Such materials may consist of the primary product as well as by-products such as manufacturing waste. These materials may undergo transport and transformation processes, which in turn may yield secondary substances.

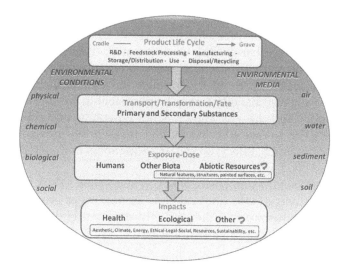

FIGURE 7.1
Comprehensive environmental assessment (CEA) framework. (*Source*: U.S. EPA 2011a.)

Note that these and subsequent events occur in various environmental media and conditions. *Environmental media* include air, water, sediment, and soil; *environmental conditions* include variables that could affect the fate of these primary and secondary substances, namely physical (e.g., temperature, wind/water currents), chemical (e.g., pH, reactions with other chemicals in the environment), biological (e.g., natural organic matter, bacteria), and social (e.g., human behavior patterns) factors. Environmental media and conditions are relevant throughout the chain of events encompassed by the CEA framework, including the potential for exposure to primary and secondary substances.

We refer to the *potential for exposure* because the mere existence of a substance in the environment does not necessarily imply that an organism or receptor of interest will be in contact with it. Exposure implies contact between a substance and a receptor, either living (i.e., humans and other biota) or nonliving (e.g., structures, landscapes). Moreover, because contact (exposure) does not necessarily mean that the substance will enter a receptor, we distinguish *dose* from exposure. Dose implies that a substance has crossed a biological or other barrier (e.g., surface layer) and been taken up by a receptor. Exposure and dose can be *cumulative* (involving multiple associated substances) and/or *aggregate* (involving a single substance via multiple pathways).

Ultimately, CEA considers the type and magnitude of both direct and indirect effects on various receptors and other types of impacts in relation to exposure–dose potential by all relevant pathways. It is important to realize that there are always multiple "impacts," some of which may be deemed desirable and others of which may be deemed undesirable, depending on specific circumstances or a given individual's perspective.

7.1.1.2 CEA as Process

Compiling the previously cited information in the CEA framework requires searching and interpreting the scientific literature as well as other sources of information. It is a fundamental step in the CEA *process*, outlined in Figure 7.2, but only an initial step. The CEA process uses the information gathered in the CEA framework as a starting point to reach transparent judgments about the implications of such information. The process accomplishes this using a collective judgment procedure that incorporates diverse perspectives of various technical experts and stakeholders. This process includes evaluation of: (1) the scope of the CEA,* (2) the completeness and accuracy of information in the framework document, and (3) the implications of information compiled in a CEA framework. The point of this is not simply to identify every conceivable concern pertaining to an issue but, as indicated in Figure 7.2, to determine *priorities* among information gaps (for research planning) and *risk trade-offs* (for risk management).

The state of the science or knowledge base largely determines which of the two paths in Figure 7.2 can be taken and, hence, what types of input or guidance the CEA process can provide to decision makers and managers. For issues such as nanotechnology and other emerging technologies, which may have relatively limited information available, it may only be feasible to provide guidance to researchers and research planners in deciding which information gaps to address. For issues that have a more developed knowledge base, the CEA process can provide guidance to risk managers in deciding which risks and benefits may warrant their highest attention. As shown in Figure 7.2, both pathways can lead to outcomes (e.g., research results, monitoring data) that would, over time, provide feedback to the CEA framework. This iterative process is consistent with the concept of adaptive management (Wintle and Lindenmayer 2008).

Collective judgment based on diverse perspectives is a key feature of the CEA process. Not only is there reason to think that some types of judgments are improved by involving a wider range of perspectives than would be provided by a small set of specialized experts (Page 2007), such a process is also consistent with greater transparency in an assessment.

---

* Note that the scope of a CEA is subject to open review along with the completeness and accuracy of information compiled in the CEA framework document. Even a comprehensive assessment cannot address an infinite number of considerations, but a CEA seeks to avoid arbitrarily setting boundaries that could omit significant impacts or other considerations. It does this in part by engaging a broad set of independent reviewers who can point out what may have been omitted from the scope of the assessment.

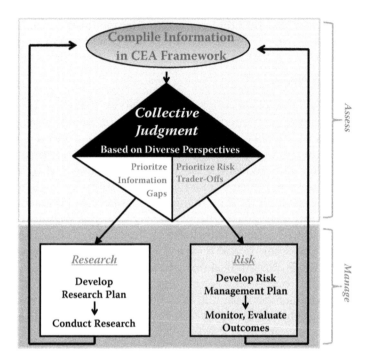

FIGURE 7.2
Comprehensive environmental assessment (CEA) process. (*Source*: U.S. EPA 2011a.)

As to how to do this, it is critically important to employ a structured procedure that allows all participants to have an equal opportunity to contribute their views. This contrasts with the more common free-for-all discussions of most workshops that, even when moderated by a designated leader or facilitator, are often dominated by a small number of outspoken individuals.

Several approaches are available for structuring a collective judgment process. Nominal group technique (NGT), for example, has been used to identify and prioritize information gaps for selected U.S.EPA nanomaterial case studies (U.S.EPA 2010a). NGT provides a relatively simple way for a set of individuals to consider and select their preferred choice(s) from a number of alternatives or options (Delbecq and Van de Ven 1971; Sink 1983). Other methods are also available, including multicriteria decision analysis (MCDA) (Belton and Stewart 2001), expert elicitation (Otway and von Winterfeldt 1992), and variations on these and other approaches (Linkov and Seager 2011; Stahl et al. 2011).

The choice of method will depend in part on the state of the science related to an issue. For issues that have a more mature knowledge base, techniques such as MCDA and expert elicitation may be more suitable, whereas a less analytic or quantitative approach, such as NGT or Delphi (Linstone and Turoff 1975), may be more appropriate for nanotechnology or other issues that are relatively lacking in empirical data. In addition, practical constraints (e.g., time, funding) may favor some methods over others that can be rather labor-intensive. Whatever the choice of method, it is important to apply it carefully to avoid making assumptions that can lead to premature structuring of the judgment process, which is especially a concern with newly emerging or data-limited issues, where unknowns may far outnumber what is known about an issue.

7.1.1.3 CEA as Meta-Assessment

As meta-assessment, CEA builds on other assessment approaches, including LCA, health risk assessment, ecological risk assessment, cost–benefit analysis, and decision science methods. Like LCA, for example, CEA incorporates a product life cycle perspective, but it also differs from most LCAs by including environmental fate, exposure-dose, human health, ecological, and other impacts as equal, integral components of the basic CEA framework (Figure 7.1) rather than as modifications or add-on features.

Moreover, whereas risk assessment (RA) and LCA typically focus on, if not require, quantitative data and analyses, CEA gives consideration to both quantitative and qualitative information. In characterizing the risks and benefits associated with an issue, a CEA might include, for example, quantitative health reference values for a given chemical while using qualitative or ordinal estimations of the potential releases of the chemical. Even though a quantitative risk estimate might not be feasible in such cases, CEA could still provide useful guidance to risk managers in weighing options to pursue.

Another sense in which CEA is a meta-assessment is that it includes results of prior assessments and analyses and, through a collective judgment process, enables seemingly contradictory or inconsistent results to be weighed and resolved in reaching an overall judgment.

Given that the goal of CEA is to provide risk managers a clear sense of the most important trade-offs related to a technology option, CEA also shares the inherently comparative nature of cost–benefit analysis and of many LCAs. However, it is also important to note that CEA does not merely provide points of comparison but identifies *priorities* among various trade-offs. Many assessments and research plans may point to a multitude of concerns or possible actions, but CEA provides a means of prioritizing through a collective judgment process for both research planning and risk management purposes (Figure 7.2).

7.1.1.4 CEA and Transparency

Transparency might be defined in various ways, but when applied to CEA or assessments generally, we mean an explicit account of how judgments and conclusions are reached. In contrast, most LCAs and RAs are prepared by, at most, a small team of specialists who use professional judgment and other factors in arriving at their conclusions. The basis for determining which factors are weighted more heavily, or how assumptions are made or justified, is seldom described in enough detail to enable an independent observer or reviewer to follow the assessors' train of logic. Such an assessment may be subjected to peer review and public comment, but these review processes do not necessarily open up the black box of the assessment process.

CEA is a transparent assessment approach because it can provide a record that would enable an independent observer (e.g., a scientific historian) to understand at a later time how the conclusions of the assessment were reached. A CEA "record" exists not only in the framework document as a catalogue of information considered for the assessment, but in the collective judgment aspect of the CEA process. By using a formal procedure that provides an explicit record of key facts, values, and other information cited by the participants themselves and, even more importantly, an objective measure of the participants' collective judgment of priorities, such as a rank-order list derived through a multivoting process, the CEA process makes the resulting assessment more transparent.

7.1.2 Illustration of CEA Applied to Selected Nanomaterials

A series of case studies of selected products incorporating nanomaterials has been developed by the U.S.EPA. These case studies have focused on nanoscale titanium dioxide ($nTiO_2$) used for water treatment and for topical sunscreen (U.S.EPA 2010c) and on nanoscale silver used in disinfectant spray (U.S.EPA 2010b). Another case study is in preparation on carbon nanotubes used in flame-retardant textiles (C. Powers, pers. comm.). These case study documents were structured around the CEA framework and included chapters on product life cycle stages, environmental fate (transport and transformation), exposure-dose, and ecological and human health effects. As noted previously, CEA framework documents serve as a foundation for the CEA process, which in these instances took the form of collective-judgment workshops. One of the case studies, nanoscale titanium dioxide ($nTiO_2$), will be highlighted in the following sections to illustrate how CEA was applied to identify and prioritize information gaps that would need to be filled in order to support future assessments of $nTiO_2$.

7.1.2.1 Nanoscale Titanium Dioxide Case Study: CEA Framework

The importance of the product life cycle is quickly evident in considering the potential impacts of a nanomaterial such as $nTiO_2$, which is used in numerous applications ranging from coatings to water treatment agents and in closed industrial settings to general consumer products. The opportunities for exposure to $nTiO_2$ are likely to be quite different, depending on whether or not the substance is tightly bound in a matrix. For example, $nTiO_2$ used in light-emitting diodes would appear to pose less potential for dispersion in the environment than $nTiO_2$ used as a water treatment agent. As a water treatment agent, there could be several opportunities for a powder of nanoscale particles to be released to the environment subsequent to manufacturing, including spillage during distribution, storage, and use. In addition, differences in manufacturing processes have been found to yield different physical and even toxicological properties of nominally equivalent nanomaterials (Dreher 2004). Thus, to evaluate the full range of potential ecological and health impacts associated with any given nanomaterial, it is necessary to consider the broader life cycle context for the material in question.

Using water treatment applications of $nTiO_2$ as an example, the product life cycle begins with the feedstocks from which the material is produced. Either titanium chloride or titanium sulfate can serve as feedstocks for producing $nTiO_2$, with the possibility of some contamination of the end product related to these respective compounds (e.g., chlorine contamination of $nTiO_2$ produced from $TiCl_4$). As part of a CEA, one would want to consider the potential for environmental releases of contaminants related to feedstock procurement and processing. Although this may not necessarily pose a significant issue in the case of feedstocks for $nTiO_2$, it is conceivable that other types of nanomaterials such as cadmium (e.g., in quantum dots) could be more problematic in this regard. This would depend in part, however, on the magnitude of feedstock use for nanoscale material production. For example, if the mass of nanomaterial-related feedstock is trivially small in relation to use of the same feedstock for bulk products, then the differential in environmental contamination from the feedstock for nanomaterial production would presumably be correspondingly small.

Manufacturing of $nTiO_2$ may be accomplished by various processes, including hydrolysis of a sol-gel (a solution of suspended colloids that forms a gel) or solution of titanium sulfate or, for larger scale production, chemical vapor deposition. The latter may in turn involve a variety of methods for vapor generation, but whether these different methods yield different physical or toxicological properties is unknown. Postproduction processing of the materials, e.g., through use of sonication, a technique using ultrasound waves, or surfactants to achieve or maintain nanoscale properties of the particles, could introduce yet another variable affecting the characteristics of the end product. Although worker exposure to a nanoscale product is the most salient concern, whether by inhalation, dermal absorption, or ingestion (e.g., resulting from hand-to-mouth activity), exposure to waste by-products associated with the manufacturing process should also be considered as part of a CEA evaluation. In addition, releases of material, both the primary product and waste by-products, outside the confines of a manufacturing facility need to be included in the scope of a CEA.

Distribution of the manufactured product involves packaging and transportation of the material. In the case of $nTiO_2$ used for water treatment, it appears that one commercial form of the product may be shipped as a powder in 10-kg "multilayer ventilated paper bags, equipped with an additional polyethylene lining when required" (Evonik 2012). This raises questions about the potential for accidental as well as routine spillage during packaging and subsequent transport of the material, with implications for workplace as well as broader environmental contamination. Similar issues would apply to product storage, with added concerns about the breach of packaging or containment material by vermin. The latter scenario would have possible relevance to wider environmental contamination through introduction of the material into the food web.

Nano-TiO_2 can be used in various ways as a water purification agent, e.g., to inactivate bacteria or as a means to remove arsenic from water by converting arsenite [As(III)] to arsenate [As(V)]. These differing uses could have different implications for releases to the environment. However, assuming that the product is mixed with water as a slurry (other scenarios are possible), one could envision the release of particles to air in the microenvironment as the powder is being prepared for mixing and/or is actually being mixed with water. After a slurry is formed, the particles could behave in various ways, but assuming the particles are not destroyed by the water treatment process itself, some portion of the particles might remain in solution in the treated water. Another possibility is that a portion of the $nTiO_2$ could settle with floc (the suspended water treatment chemicals) in the sedimentation stage of water treatment and be subject to removal as sludge.

The disposal of sludge created in the water treatment process could follow several environmental pathways, including landfills and land applications. The latter conjures scenarios such as application to land used for growing crops, grazing animals, recreational uses such as parks, and numerous other uses that could pose direct and indirect opportunities for exposure of humans and other biota. Transport and transformation processes could also come into play through surface runoff, plant uptake, and a host of other conceivable events.

The previous discussion highlights some examples of points that warrant consideration in a CEA of nanomaterials, but it in no way does justice to the complexity of the exposure component of such an assessment. For example, it is important to recognize that exposure may be both cumulative and aggregative. *Cumulative exposure* refers to the multiple contaminants, including waste by-products and secondary transformation products, that could be associated with a given nanomaterial such as $nTiO_2$. *Aggregate exposure* refers to the multiple environmental sources, pathways, and routes through which exposure to a nanomaterial might occur. For example, given that $nTiO_2$ may be found in various consumer products such as sunscreen lotions, cosmetics, foodstuffs, etc., any exposure to $nTiO_2$ in connection with its use as a water treatment agent should be understood in relation to the total potential exposure to $nTiO_2$ across sources, pathways, and routes. Further complexities arise when time and activity patterns of exposed organisms are considered.

Exposure characterization provides a context and premise for considering the effects of nanomaterials on both ecological receptors and human populations, for without exposure there can be no effects. As discussed in Chapter 5, with regard to ecological effects, some studies using standard testing assays indicate that $nTiO_2$ may be toxic to water fleas (*Daphnia magna*), a key aquatic indicator species (Lovern and Klaper 2006; Wiench et al. 2007). Also, $nTiO_2$ has bactericidal properties (Coleman et al. 2005; Rincon and Pulgarin 2003; Kuhn et al. 2003), which may be desirable under controlled conditions but undesirable if beneficial bacteria in the environment are affected. Such effects may be modulated by various factors, including particle size (Hund-Rinke and Simon 2006) and material preparation (Lovern and Klaper 2006). It also appears that $nTiO_2$ can affect the uptake of other substances. As described in Chapter 5, Sun et al. (2007) found that As(V) strongly binds to $nTiO_2$ in water and that the presence of $nTiO_2$ more than doubles the uptake of arsenic in carp. Similar findings were reported by Zhang et al. (2006) for cadmium uptake in carp in the presence of $nTiO_2$. Although toxicity was not assessed in these studies, the increases in metal uptake suggest that interactive/secondary effects warrant careful attention as part of a CEA of such nanomaterials.

Information on the health effects of $nTiO_2$ is not as plentiful as one might prefer, but it is growing and can only be highlighted here to make a few general points. A key point is that extrapolation from bulk or microscale TiO_2 to $nTiO_2$ is inadvisable, given the notable differences in physicochemical properties of nanoscale and microscale TiO_2. Oberdörster, Ferin, and Lehner (1994) observed differences in particle retention, translocation, pulmonary inflammation, and impairment of alveolar macrophage function between nanoscale (ultrafine) and microscale (fine) particles of TiO_2 after twelve weeks of inhalation exposure in rats when compared on the basis of the mass of the dose. However, when compared in terms of total particle surface area (given that $nTiO_2$ has a greater surface area per mass than microscale TiO_2), a linear dose–response curve was apparent for the $nTiO_2$. Other studies have demonstrated that surface area may account for differences in respiratory toxicity effects between nanoscale and microscale TiO_2 (e.g., Bermudez et al. 2004; Warheit et al. 2007). However, other factors, including surface coatings or contamination, surface charge, and primary particle size, may also contribute to toxic properties of $nTiO_2$ (Warheit et al. 2007; Kreyling et al. 2002). In addition, some high-dose respiratory effects in rats may have been confounded by particle overload due to species differences in lung clearance mechanisms and thus not be representative of effects in humans under occupational or general environmental exposure conditions (Bermudez et al. 2002, 2004).

Data for other target organs are quite limited, especially for reproductive, developmental, and immunological endpoints. However, some information indicates that nanoparticles such as $nTiO_2$ may cross the blood–brain barrier, be taken up in the brain, and induce certain effects in brain cells (microglia), at least *in vitro* (Long et al. 2006, 2007). In some cases, transport to the brain may occur directly via the olfactory nerve (Oberdörster et al. 2004). As with other types of nanoparticles, oxidative damage appears to be a common mechanism of toxicity associated with $nTiO_2$ (Long et al. 2006, 2007; Xia et al. 2006).

At present, the available data do not appear to be sufficient to derive quantitative hazard assessments for $nTiO_2$ or for nanomaterials in general. However, the previously discussed highlights of the effects information for both ecological receptors and experimental animal subjects suggest that assessments may soon be feasible if research is properly targeted, which was in fact the reason for preparing the U.S.EPA nanomaterial case studies and for holding workshops to identify and prioritize research directions for these materials.

7.1.2.2 Nanoscale Titanium Dioxide Case Study Workshop: CEA Process

Given that the knowledge base for $nTiO_2$ was not sufficiently developed to support an actual assessment, the CEA process for $nTiO_2$ followed the path toward research planning shown in Figure 7.2. The collective-judgment workshop for $nTiO_2$ used the framework document described previously as a starting point, with invited participants being required to review the document in advance of the two-day workshop. Approximately 50 individuals were selected to represent a range of technical fields (e.g., manufacturing, environmental fate, exposure, ecology, toxicology, risk management) and sectors (academic, civic, government, industry).

An NGT process was used for the meeting, which involved a formal round-robin procedure (U.S.EPA 2010a; Delbecq and Van de Ven 1971) having each participant state the issue(s) they considered the most important to address in order to conduct an assessment of $nTiO_2$ using the CEA approach. Given practical constraints of time and attention with such a large group, the participants were divided into two groups of approximately 25 each for the round-robin segment. The NGT format allowed participants to hear a variety of viewpoints about key information gaps and thereby obtain a broader perspective of the issues than that of their own particular expertise.

Each subgroup then consolidated closely related or overlapping issues and voted on these topics using a multivoting procedure. By allowing each person ten votes, weighted from 1 to 10, a ranked order of priorities was obtained for each subgroup. Subsequently, the two subgroups were brought together to consolidate their respective priorities and have another multivote to produce an overall list of rank-order priorities. In this way, the process provided a collective judgment about which information gaps warranted the most attention to support a future CEA of $nTiO_2$.

A detailed account of the $nTiO_2$ case study workshop may be consulted for more specific descriptions of the findings that resulted from the collective judgment process (U.S.EPA 2010a), but in general terms, the group prioritized the improvement of methods to characterize $nTiO_2$, the development of exposure scenarios, better understanding of environmental fate processes, and more studies of toxicity in humans and other biota. Specific areas within these themes included: (1) developing test methods for ecological and human health effects, (2) characterizing $nTiO_2$ physicochemical characteristics throughout the product life cycle, and (3) determining mechanisms of action for $nTiO_2$. It should be noted that the workshop process resulted in the participants identifying a number of research needs not originally singled out in the case study document. Examples of these needs include: (1) determining criteria for harmonized test protocols, (2) measuring background levels of $nTiO_2$ in humans, and (3) evaluating how analytical methods to characterize mass flow and other risk factors can be used in comparing environmental trade-offs of $nTiO_2$ use.

While some of these research needs specifically relate to $nTiO_2$ research, others apply more broadly to nanomaterials as a whole. This was expected, as the explanatory background materials provided to the participants noted that they should consider which research needs would apply to $nTiO_2$ in water treatment alone, or to $nTiO_2$ regardless of application, or to nanomaterials more generally (i.e., not only to $nTiO_2$). Subsequent case study workshops focusing on other nanomaterials and applications have pointed to some of the same general concerns as well as other priorities more specific to the particular nanomaterial (U.S.EPA 2011b).

7.2 Other Life Cycle-Based Assessment Approaches

In addition to Nano LCRA and CEA, two other notable examples of the idea of incorporating the product life cycle in assessing nanomaterials warrant discussion here. One is the NanoRisk Framework jointly developed by the Environmental Defense Fund and DuPont (ED DuPont 2007), and the other is the RA-LCA-MCDA approach described by Linkov and Seager (2011). After highlighting key features of these approaches, some comparisons of the four examples will be offered.

7.2.1 The NanoRisk Framework and LC-RA-MCDA Approach

The ED-DuPont NanoRisk framework is intended to help organize what is known; assess, prioritize, and address data needs; and communicate how risks are managed (ED DuPont 2007). The framework is meant to be holistic by explicitly including the product life cycle consisting of material sources, production, use, and end-of-life disposal/recycling.

A key feature of the NanoRisk framework is the development of informational profiles or "base sets" about the properties, hazards, and exposures associated with a given nanomaterial and its application. These base sets are the foundation for the multistep process used in implementing the framework (ED DuPont 2007):

1. Describing the material and its application
2. Profiling the material life cycle in terms of properties, potential safety, health and environmental hazards, and opportunities for human or environmental exposure at each step of the product life cycle
3. Evaluating risks, either with available data or by assuming the "reasonable worst case"
4. Assessing risk management options, including engineering controls, protective equipment, risk communications, and process or product modification
5. Decide, document, and act

The NanoRisk framework formed the basis for an ISO Technical Report (ISO 2011), which adds a sixth step, namely, review and adapt. The latter step appears to refer implicitly to the idea of adaptive management (see Chapter 6).

Another important feature of the NanoRisk framework is the inclusion of an "output worksheet," which is said to "facilitate evaluation, management, and communication" by providing a template for organizing information for the framework, capturing overall evaluations of that information, and recording consequent management decisions (ED DuPont 2007).

Three case studies using the NanoRisk framework were attempted (http://www.nanoriskframework.com/case-studies/), although only one (DuPont Light Stabilizer 210, a surface-treated high-rutile phase to $nTiO_2$) was completed. Full base sets of data for case studies of carbon nanotubes and nano-sized zerovalent iron (nano-FeO) were not feasible at the time, but the framework was used to identify key uncertainties regarding nano-FeO that would need to be addressed before a full evaluation could be undertaken.

Linkov and Seager (2011) have described an approach to assessing nanomaterials and other emerging technologies and "threats" using a combination of multicriteria decision analysis, life cycle assessment, and risk assessment (RA-LCA-MCDA). The addition of MCDA to the synthesis of RA and LCA is a key feature of their approach and will be highlighted here.

Linkov and Seager (2011) refer to MCDA as a group of methods used to improve understanding of a complicated or uncertain decision-making process. They describe four steps for the MCDA process:

> (1) structuring the problem by identifying criteria through stakeholder elicitation and assessment of the different criteria that are relevant to the given decision; (2) eliciting the parameters of the model, such as alternatives, decision criteria, relative weights, and preference thresholds, and evaluating the performance of each alternative on each criterion; (3) applying a decision algorithm that ranks each alternative from most to least preferred; (4) interpreting results of the model and reiterating the process from step 1 or 2 by re-evaluating the model.

The role of MCDA in their approach is to complement LCA and RA by providing a means to "overcome the human cognitive limitations associated with reasoning under high levels of uncertainty and juggling multiple decision criteria and alternatives simultaneously" (Linkov and Seager 2011). This would help those involved in the assessment and management of risks to select or prioritize among alternative issues or courses of action.

An example of the RA-LCA-MCDA approach cited by Linkov and Seager (2011) is an assessment focused on synthesis pathways for production of single-wall carbon nanotubes (SWCNT) reported by Canis, Linkov, and Seager (2010). Linkov and Seager (2011) noted that the assessment showed that the preferred choice of manufacturing process depends upon the values placed on the criteria chosen for impact evaluation, e.g., material efficiency, energy consumption, life cycle impacts, production costs, and nano-related health risks. As one might expect, different individuals might weigh various criteria differently, e.g., manufacturers might weigh efficiency aspects highly, whereas environmentalists might place greater emphasis on health risks or protection of natural resources.

7.2.2 Comparisons of Life Cycle-Based Assessment Approaches

All of the approaches described here share a basic feature: the incorporation of a product life cycle perspective as part of an assessment methodology. It is also fair to say that all of these methods are intended to be holistic and systematic, although defining precisely how successfully these objectives are met could be difficult. A detailed comparison of these approaches is beyond the scope of this chapter, but a few salient points can be noted.

The LCRA approach described in Chapter 6 is presented as a "screening-level framework," whereas all of the other three approaches described in this chapter would generally involve more time, effort, and resources. In some circumstances, a screening-level evaluation may be all that is feasible or appropriate, whereas in other situations, a more extensive assessment may be needed.

If more than a screening evaluation is warranted, the choice between CEA, NanoRisk framework, and RA-LCA-MCDA involves various factors. Both CEA and RA-LCA-MCDA incorporate a formal decision process component that is not included in the NanoRisk framework (and LCRA). Among other things, the inclusion of an explicit process to prioritize risks and/or research needs can provide greater transparency to the assessment as well as other potential benefits related to collective judgments.

The benefits of collective judgment depend in part on the nature of the group that is charged with making such judgments. The CEA approach emphasizes the importance of using a diverse and balanced group of experts and stakeholders in the collective-judgment process. Although the description of RA-LCA-MCDA alludes to the participation of "decision makers, stakeholders, or public groups" and cites an example involving manufacturers, consumers, regulators, and environmentalists (Linkov and Seager 2011), it is not clear that achieving either breadth or depth of perspectives is important to this approach, at least not to the extent that it is with CEA.

It should also be noted that CEA is not limited to one particular method for reaching a collective judgment, as is RA-LCA-MCDA. Although CEA might employ MCDA in some situations, it is not restricted to this tool alone. As described previously, nominal group technique (NGT) was used in the CEA workshops on the U.S.EPA nanomaterial case studies because NGT seemed more suited to the relatively nascent state of the science for assessing the risks and benefits of nanomaterials. However, more analytic methods, such as MCDA and expert elicitation, could also be used as part of a CEA process, especially to address issues with a more extensive and better developed knowledge base.

Much more could be said about the similarities and differences among the approaches discussed in this chapter, each of which has its respective advantages and disadvantages. And, in fact, comparison of these methods has been the subject of other discussions (Grieger et al. 2011; SRA 2010). As a clearer picture emerges of the strengths of various approaches, it seems likely that the field of risk assessment will adopt the best features of each approach.

References

Belton, V., and T. J. Stewart. 2001. *Multiple Criteria Decision Analysis: An Integrated Approach.* New York: Springer-Verlag.

Bermudez, E., J. B. Mangum, B. Asgharian, B. A. Wong, E. E. Reverdy, D. B. Janszen, P. M. Hext, D. B. Warheit, and J. I. Everitt. 2002. Long-term pulmonary responses of three laboratory rodent species to subchronic inhalation of pigmentary titanium dioxide particles. *Toxicol. Sci.* 70(1): 86–97.

Bermudez, E., J. B. Mangum, B. A. Wong, B. Asgharian, P. M. Hext, D. B. Warheit, and J. I. Everitt. 2004. Pulmonary responses of mice, rats, and hamsters to subchronic inhalation of ultrafine titanium dioxide particles. *Toxicol. Sci.* 77(2): 347–57.

Canis, L., I. Linkov, and T. P. Seager. 2010. Application of stochastic multiattribute analysis to assessment of single walled carbon nanotube synthesis processes. *Env. Sci. Technol.* 44:8704–11.

Coleman, H. M., C. P. Marquis, J. A. Scott, S.-S. Chin, and R. Amal. 2005. Bactericidal effects of titanium dioxide-based photocatalysts. *Chem. Eng. J.* 113:55–63.

Cordis. 2004. Towards a European strategy for nanotechnology. *COM* 2004: 338. http://cordis.europa.eu/nanotechnology/actionplan.htm

Davis, J. M. 2007. How to assess the risks of nanotechnology: Learning from past experience. *J. Nanosci. Nanotech.* 7:402–9.

Davis, J. M., and W. H. Farland. 2007. The paradoxes of MTBE. *Toxicol. Sci.* 61:211–17.

Davis, J. M., and V. M. Thomas. 2006. Systematic approach to evaluating trade-offs among fuel options: The lessons of MTBE. *Ann. N.Y. Acad. Sci.* 1076:498–515.

Delbecq, A. L., and A. H. Van de Ven. 1971. A group process model for problem identification and program planning. *J. Appl. Behav. Sci.* 7:466–92.

Dreher, K. 2004. Health and environmental impact of nanotechnology: Toxicological assessment of manufactured nanoparticles. *Toxicol. Sci.* 77:3–5.

ED DuPont. 2007. Nano risk framework. http://www.nanoriskframework.org/page.cfm?tagID=1095

Evonik. 2012. AEROSIL® fumed silica—packaging. http://www.aerosil.com/product/aerosil/en/services/packaging/pages/default.aspx

Grieger, K. D., I. Linkov, S. F. Hansen, and A. Baun. 2011. Environmental risk analysis for nanomaterials: Review and evaluation of frameworks. *Nanotoxicology* 6(2): 196–212. http://dx.doi.org/10.3109/17435390.2011.569095

Hund-Rinke, K., and M. Simon. 2006. Ecotoxicological effects of photocatalytic active nanoparticles TiO_2 on algae and daphnids. *Environ. Sci. Pollut. Res.* 13(4): 225–32.

ISO. 2011. *Nanotechnologies: Nanomaterial risk evaluation,* 1st ed. Technical Report ISO/TR 13121. Geneva, Switzerland: International Standards Organization.

Kreyling, W. G., M. Semmler, F. Erbe, P. Mayer, S. Takenaka, and H. Schultz. 2002. Translocation of ultrafine insoluble iridium particles from lung epithelium to extrapulmonary organs is size dependent but low. *J. Toxicol. Environ. Health A* 65(20): 1513–30.

Kuhn, K. P., I. F. Chaberny, K. Massholder, M. Stickler, V. W. Benz, H. G. Sonntag, and L. Erdinger. 2003. Disinfection of surfaces by photocatalytic oxidation with titanium dioxide and UVA light. *Chemosphere* 53(1): 71–77.

Linkov, I., and T. P. Seager. 2011. Coupling multi-criteria decision analysis, life-cycle assessment, and risk assessment for emerging threats. *Environ. Sci. Technol.* 45:5068–74.

Linstone, H. A., and M. Turoff. 1975. *The Delphi Method: Techniques and Applications.* Reading, MA: Addison-Wesley.

Long, T. C., N. Saleh, R. D. Tilton, G. V. Lowry, and B. Veronisi. 2006. Titanium dioxide (P25) produces reactive oxygen species in immortalized brain microglia (BV2): Implications for nanoparticle neurotoxicity. *Environ. Sci. Toxicol.* 40:4346–52.

Long, T. C., J. Tajuba, P. Sama, P. Gillespie, N. Saleh, J. Parker, C. Swartz, and B. Veronisi. 2007. Nanosize TiO_2 stimulates reactive oxygen species in brain microglia and damages neurons in vitro. Abstract No. 1387. *Itinerary Planner.* Charlotte, NC: Society of Toxicology.

Lovern, S. B., and R. Klaper. 2006. *Daphnia magna* mortality when exposed to titanium dioxide and fullerene (C_{60}) nanoparticles. *Environ. Toxicol. Chem.* 25(4): 1132–37.

Oberdörster, G., J. Ferin, and B. E. Lehner. 1994. Correlation between particle size, in vivo particle persistence, and lung injury. *Environ. Health Perspect.* 102(6): 173–79.

Otway, H., and D. von Winterfeldt. 1992. Expert judgment in risk analysis and management: Process, context, and pitfalls. *Risk Anal.* 12(1): 83–93.

Page, S. E. 2007. *The Difference: How the Power of Diversity Creates Better Groups, Firms, Schools, and Societies.* Princeton, NJ: Princeton Univ. Press.

Rincon, A. G., and C. Pulgarin. 2003. Photocatalytic inactivation of *E. coli*: Effect of (continuous-intermittent) light intensity and of (suspended-fixed) TiO_2 concentration. *Appl. Catalysis B-Environ.* 44:263–84.

Sass, J. 2007. Nanotechnology's invisible threat: Small science, big consequences. NRDC issue paper. www.nrdc.org/health/science/nano/nano.pdf

Shatkin, J. A. 2005. Risk assessment: Informing the development of beneficial technology. Presented at the Advancing Beneficial Nanotechnology Foresight Conference, San Francisco. http://www.foresight.org/conference2005/presentations/shatkin.pdf

Sink, D. S. 1983. Using the nominal group technique effectively. *National Productivity Review* 2(2): 173–84.

SRA. 2010. Society for Risk Analysis Symposium: Decision Support Methods for Nanomaterial Risk Assessment and Risk Management, Salt Lake City, UT.

Stahl, C., A. Cimorelli, C. Mazzarella, and B. Jenkins. 2011. Toward sustainability: A case study demonstrating trans-disciplinary learning through the selection and use of indicators in a decision making process. *Integr. Environ. Assess. Manage.* 7(3): 483–98.

Sun, H., X. Zhang, O. Niu, Y. Chen, and J. C. Crittenden. 2007. Enhanced accumulation of arsenate in carp in the presence of titanium dioxide nanoparticles. *Water Air Soil Pollut.* 178:245–54.

Sweet, L., and B. Strohm. 2006. Nanotechnology: Life-cycle risk management. *Human Ecol. Risk Assess.* 12(3): 528–51.

U.S. EPA. 1992. Alternative fuels research strategy. External Review Draft. EPA 600/AP-92.002. http://www.epa.gov/ncea/pdfs/mtbe/altfuel.pdf

———. 1998. Research strategy for oxygenates in water. *Federal Register* 64(35): 8817.

———. 1999. Achieving clean air and clean water: The report of the Blue Ribbon Panel on Oxygenates in Gasoline. Report No. EPA420-R-99-021. http://www.epa.gov/otaq/consumer/fuels/oxypanel/r99021.pdf

———. 2007. Nanotechnology white paper. Report No. EPA 100/B-07/001. Washington, DC: EPA. http://www.epa.gov/OSA/pdfs/nanotech/epa-nanotechnology-whitepaper-0207.pdf

———. 2010a. Nanomaterial case studies workshop: Developing a comprehensive environmental assessment research strategy for nanoscale titanium dioxide. Report No. EPA/600/R-10/042. Research Triangle Park, NC: EPA. http://www.epa.gov/osp/bosc/pdf/nano1005summ.pdf

———. 2010b. Nanomaterial case study: Nanoscale silver in disinfectant spray. External review draft. Report No. EPA/600/R-10/081. Washington, DC: EPA. http://cfpub.epa.gov/ncea/cfm/recordisplay.cfm?deid=226723

———. 2010c. Nanomaterial case studies: Nanoscale titanium dioxide in water treatment and topical sunscreen. Final Report No. EPA/600/R-09/057F. Research Triangle Park, NC: EPA. http://cfpub.epa.gov/ncea/cfm/recordisplay.cfm?deid=230972

———. 2011a. Comprehensive environmental assessment: A meta-assessment approach to increase effectiveness of risk management and research planning. http://www.epa.gov/nanoscience/files/CEAPrecis.pdf

———. 2011b. Nanomaterial Case Study Workshop: Developing a comprehensive environmental assessment research strategy for nanoscale silver, Research Triangle Park, NC.

Warheit, D. B., R. A. Hoke, C. Finlay, E. M. Donner, K. L. Reed, and C. M. Sayes. 2007. Development of a base set of toxicity tests using ultrafine TiO_2 particles as a component of nanoparticle risk management. *Toxicol. Lett.* 171(3): 99–110.

Wiench, K., R. Landsiedel, S. Zok, V. Hisgen, K. Radke, and B. van Ravenswaay. 2007. Aquatic fate and toxicity of nanoparticles: Agglomeration, sedimentation and effects on *Daphnia magna*. Abstract No. 1384. *2007 Itinerary Planner*. Charlotte, NC: Society of Toxicology.

Wintle, B. A., and D. B. Lindenmayer. 2008. Adaptive risk management for certifiably sustainable forestry. *For. Ecol. Manage.* 256(6): 1311–19.

Xia, T., M. Kovochich, J. Brant, M. Hotze, J. Sempf, T. Oberley, C. Sioutas, J. I. Yeh, M. R. Wiesner, and A. E. Nel. 2006. Comparison of the abilities of ambient and manufactured nanoparticles to induce cellular toxicity according to an oxidative stress paradigm. *Nano Lett.* 6(8): 1794–807.

Zhang, X., H. Sun, Z. Zhang, Q. Niu, Y. Chen, and J. C. Crittenden. 2007. Enhanced bioaccumulation of cadmium in carp in the presence of titanium dioxide nanoparticles. *Chemosphere* 67(1): 160–66.

Additional References

Fiksel, J. 2003. Designing resilient, sustainable systems. *Environ. Sci. Technol.* 37(23): 5330–39.

UNEP/SETAC. 2005. *Life Cycle Approaches: The Road From Analysis to Practice.* Paris: UNEP/SETAC.

Shatkin, J. 2008. *Nanotechnology: Health and Environmental Risks (Perspectives in Nanotechnology).* Boca Raton, FL: CRC Press.

Robichaud, C. O., D. Tanzil, and M. R. Wiesner. 2007. *Assessing life cycle risks of nanomaterials.* New York: McGraw Hill.

8

Managing Risks in Occupational Environments

Thomas M. Peters

CONTENTS

This chapter focuses on efforts to protect workers from health hazards that may result from handling and managing engineered nanomaterials in occupational settings. Exposures in these settings are often many times greater than those that occur via environmental releases or those experienced by consumers of products that contain nanomaterials. Consequently, managing risks posed by exposures to engineered nanomaterials in the workplace is of critical importance.

Traditionally, professionals trained in the field of industrial hygiene have held the responsibility to assess and mitigate workplace exposures. The term *environmental health and safety* (EHS) is often used to more accurately describe the job responsibilities of modern industrial hygienists that include assessing the release of hazardous compounds from the workplace to the environment.

The U.S. workforce that handles, produces, or disposes of engineered nanomaterials is growing. An estimated $32 billion worth of products that incorporate nanomaterials were sold in the United States in 2005 (Lux Research 2006), a number that is projected to increase to $3 trillion by 2015 (Lux Research 2009). Governmental agencies, companies, and research organizations increasingly are taking proactive steps to understand and mitigate the potential adverse health consequences of engineered nanomaterials in occupational environments. Although the need to identify potential occupational hazards for nanomaterials is recognized generally, the financial impetus and commitment of resources to support such initiatives have been small compared to those directed toward nanomaterials research and development efforts. In the United States, funding for nanotechnology issues related to EHS was projected at $120 million in 2011 (NSET 2010), whereas that for research and development of nanotechnology was estimated at $18 billion (Lux Research 2009).

The risks presented by ignoring the potential hazards of nanomaterials are numerous. They include unanticipated adverse health effects and diseases from exposure to nanomaterials among workers and the general public, fears and the loss of confidence among the public regarding the use of products and materials containing nanomaterials, and finally, the financial costs of liability and litigation due to personal and environmental exposures. A proactive approach may avoid the familiar history of identifying the negative health and environmental impacts of industrial and commercial materials only after years of their extensive production, use, and release into the environment. A few notorious examples from the latter half of the twentieth century include asbestos, lead (discussed in Chapter 3), silica, and a variety of toxic solvents.

In this chapter, several issues that represent challenges for protecting workers from exposure to engineered nanomaterials in the workplace are defined. The discussion then shifts to activities central to the practice of industrial hygiene (anticipation, recognition, evaluation, and control) in the context of managing risks from exposure to nanomaterials. Also discussed are the research needs identified by various national and international organizations to help manage risks in the workplace. Lastly, the directions of current efforts to address limitations in managing risks from nanomaterial hazards in the workplace are presented. The focus of the chapter is on inhalational routes of exposure primarily because the respiratory system is a particularly susceptible organ for particulate exposures and also because the literature contains relatively little information for other routes of exposure.

8.1 Defining the Issues

8.1.1 Terminology and Classification Schemes

Knowledge of terminology facilitates effective communication and management of risk. For this reason, considerable efforts have been made to standardize the vocabulary used for engineered nanomaterials (Abe 2011). Engineered nanomaterials are materials with any external dimension in the nanoscale (<100 nm) or having internal structure or surface structure in the nanoscale. Nanomaterials can be classified as nano-objects and nanostructured materials. Nano-objects are materials with one, two, or three dimensions in the nanoscale and include nanoparticles (all external dimensions <100 nm), nanofibers (two similar external dimensions <100 nm), and nanoplates (one external dimension <100 nm). Nanostuctured materials are materials having internal nanostructure or surface nanostructure.

A typical workplace has particle contamination from numerous sources. Particles may be of natural origin, such as unwanted by-products of industrial processes, or they may be engineered for a particular purpose. An ultrafine particle is a term used to describe a particle with all dimensions smaller than 100 nm that is produced by natural sources or combustion processes outside the workplace. Ultrafine particles are found commonly in outdoor air at concentrations that exceed those in the workplace and may be brought into the workplace as dilution air in ventilation. Incidental nanoparticles are by-products of high-temperature processes, such as welding, that are produced inside the workplace. Within the context of EHS, it is important to distinguish engineered nanomaterials from background ultrafine particles, incidental nanoparticles, and other larger particles that may be present.

Engineered nanomaterials may be classified according to their bulk composition and structure. Carbon-based materials are composed mostly of carbon, commonly taking the form of hollow spheres, ellipsoids, or tubes. Metal-based materials include nanogold, nanosilver, and metal oxides. Semiconductor-based materials include quantum dots with size-dependent optical emission properties. Polymer-based nanomaterials include dendrimers, which are nanosized polymers built from branched units with numerous chain ends that can be tailored to perform specific chemical functions. Composites are engineered nanomaterials combined with other nanomaterials or with larger, bulk-type materials, such as carbon nanotubes (CNTs) added to epoxy for reinforcement.

Engineered nanomaterials may be free or fixed within the matrix of a larger structure. For example, carbon nanotubes are free when in bulk form in the container in which they were purchased and are fixed when incorporated into epoxies that are used in sporting goods or airplane wings. This distinction has an important bearing on the mobility of the nanomaterials within the workplace. Low-energy processes, such as opening a container, may aerosolize nanomaterials that are in free form, whereas much higher energies, such as sanding, are required to disperse nanomaterials if they are fixed.

8.1.2 Complexity of Nanomaterials

Each nanomaterial that is engineered to enhance products or to enable new processes represents a new workplace hazard. Engineered nanomaterials can be tuned to achieve specific properties by changing the size, shape, or the surface coating of the material. For example, certain kinds of surface functional groups can be added to the surface of carbon nanotubes to make them soluble in water, which likely also changes their bioavailability and, hence, their toxicity in human systems. This inherent uniqueness of each nanomaterial with small changes in a specific property has resulted in thousands of novel materials now in the workplace for which there are no toxicological or epidemiological data.

A given engineered nanomaterial may be present in different forms within a workplace. Engineered nanoparticles are often sold as powders that clump together to form micrometer-sized particles, or agglomerates. Some nanomaterials consist of particles that are substantially outside the nanoparticle size range but have nanoscale features that give them their unique and useful properties. The lithium titanate nanomaterial depicted in Figure 8.1 is an example. The nanomaterial is composed of micrometer-sized spherical particles (Figure 8.1a–c) with nano-sized clusters on their surface (Figure 8.1d).

Nanomaterials are also often embedded into matrices (e.g., CNTs are embedded into epoxy to strengthen a product without adding weight). Such embedded nanomaterials may become liberated or exposed when the matrix is disturbed in various manufacturing processes. For example, sanding epoxy objects that have been reinforced with CNTs emits airborne particles that are composed of epoxy cores with CNT

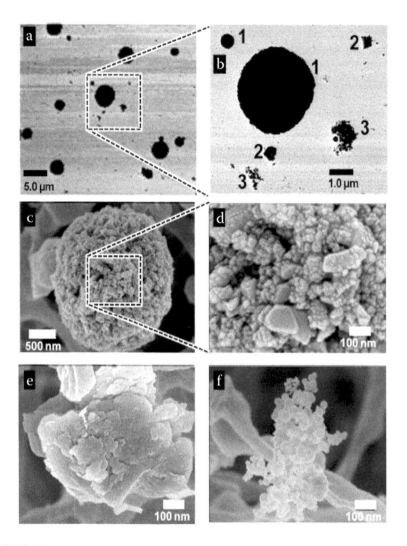

FIGURE 8.1

Electron microscopy images from filter samples depict a range of particle sizes and morphologies for particles collected in a manufacturing facility that produces high-surface-area metal oxide (lithium titanate) nanomaterials for fuel cell applications. Transmission electron micrograph images (a and b) show three types of particles of different shape and size: large spherically shaped particles (b,1), irregularly shaped particles (b,2), and smaller particle chains (b,3). The scanning electron micrograph images (c–e) reveal that the larger spherical particles are actually composed of smaller nanoparticles 10–80 nm in size interwoven into larger aggregates (c,d); irregularly shaped particles have an amorphous structure (e); and chain agglomerates are composed of spherical nodules of 5–50 nm in size. Elemental analysis of these particles shows that only the larger spherical particles are composed of titanium and therefore are due to the manufacturing of nanomaterials. (Reprinted from Peters et al. [2006] with permission of Taylor & Francis.)

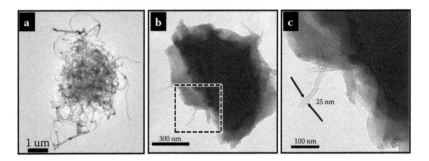

FIGURE 8.2
Electron microscopy images depict different airborne particles sampled in a facility that produces CNT-epoxy nanocomposites. The morphology of multiwalled CNTs purchased from a manufacturer are typically tangled bundles of 500 nm to several micrometers in diameter (a). Airborne particles that are emitted from sanding epoxy with CNTs embedded for strength appear as an epoxy core with protrusions (b) that have morphology consistent with CNTs when viewed at greater magnification (c). (Adapted from Cena and Peters [2011] with permission of Taylor & Francis.)

protrusions (Figure 8.2b–c), which are very different from bulk CNTs (Figure 8.2a).

These very different forms of nanomaterials all represent workplace hazards that the EHS professional needs to address. A further complication is that these nanomaterials disperse within a workplace that already has a background level of particles. In the case of the lithium titanate nanomaterial production facility, other particles were common in the workplace air (Figure 8.1b,e,f). Consequently, exposure assessment methods must be specific for the particular nanomaterial of concern in the workplace.

8.1.3 Regulations Applicable to Workplaces

Currently, there are no enforceable occupational exposure limits (OELs) specific to engineered nanomaterials. The Occupational Safety and Health Administration (OSHA) has established generic mass-based permissible exposure limits (PELs) that apply to airborne exposures in workplaces where nanomaterials are handled and produced. A PEL of 15 mg m^{-3} for respirable and 5 mg m^{-3} for total dust applies to particles not otherwise regulated (PNOR), which is based on the fact that the physical presence of biologically inert, insoluble or poorly soluble, low-toxicity particles can overload the clearance mechanisms of the respiratory system. However, these PELs for PNOR are very high and typically of little practical value for most workplaces. Composition-specific PELs are available that apply to some nanomaterials, such as silver metal (0.01 mg m^{-3} for respirable particles) and titanium dioxide (15 mg m^{-3} for total particles).

Each workplace is required to demonstrate compliance with these limits through gravimetric sampling.

8.1.4 NIOSH Current Intelligence Bulletins

As discussed in Chapter 4, particle number or surface area concentration may be metrics that are more relevant to the development of adverse health effects than mass concentration. In some cases (e.g., titanium dioxide, TiO_2), materials impart greater adverse biological behavior when they are nanostructured or nano-sized than when they occur as larger particles (Maynard, Warheit, and Philbert 2011). Also, the samplers that are used to collect samples to demonstrate compliance (e.g., respirable cyclone or closed-faced filter cassette) collect larger particles in addition to nanoparticles. The National Institute for Occupational Safety and Health (NIOSH) develops a current intelligence bulletin (CIB) to address limitations in PELs and the sampling methods that are used to show compliance to them. In the CIB, a quantitative risk assessment is presented that includes dose–response relationships derived from available animal and human data. These relationships are used to establish recommended exposure limits (RELs) and assessment strategies to demonstrate that exposures are below these levels. RELs typically represent levels that, over a working lifetime, are estimated to reduce risks to below 1 in 1,000.

In the published CIB for TiO_2 (NIOSH 2011), NIOSH states that an unknown number of U.S. workers produce and handle an estimated 1.5 million metric tons of TiO_2 that are incorporated into a wide variety of commercial products, including paints, cosmetics, and food. Some of this material is unintentionally or intentionally produced in fine or ultrafine (nanoparticles) size fractions to achieve characteristics favorable in manufacturing or product performance. Scientific evidence suggests that persistent pulmonary inflammation and lung tumors scale with the particle size and surface area concentration of TiO_2 exposures. NIOSH, therefore, sets forth recommended exposure limits (RELs) for TiO_2, depending on the size of the particles in the air: 1.5 mg m^{-3} for fine TiO_2 and 0.3 mg m^{-3} for ultrafine (including engineered nanoscale) TiO_2. These RELs are for time-weighted average concentrations for up to ten hours per day during a forty-hour work week. They further recommend that exposures be controlled as low as possible below these RELs. In this CIB, NIOSH further suggests that these adverse health effects may not be material-specific but due to a generic effect of poorly soluble, low-toxicity particles in the lung.

NIOSH has also drafted a CIB that is available for public comment for carbon nanotubes and nanofibers in the workplace (NIOSH 2010). They propose an REL of 7 µg m^{-3} (8-hour time-weighted average workshift exposure during a 40-hour work week) for carbon nanotubes and nanofibers measured as elemental carbon by NIOSH Method 5040 to prevent excess risk of pulmonary inflammation and fibrosis. The risk assessment presented in the CIB suggests that workers may have >10% excess risk of developing early-stage pulmonary fibrosis if exposed at the REL for a full working lifetime. However, the REL was set as the limit of quantification of NIOSH Method 5040, which NIOSH has selected as the best available method to assess exposures. This method is nonspecific for carbon nanotubes and nanofibers, with other sources of elemental carbon possible in workplace settings. Consequently, NIOSH encourages development of more suitable sampling and analytical methods, which may include microscopic methods like those used to assess exposure to asbestos.

8.1.5 U.S.EPA Regulations

Regulations set by the Environmental Protection Agency (U.S.EPA) have relevance to engineered nanomaterials in occupational settings. The U.S.EPA considers carbon nanotubes to be chemically different from conventional carbon compounds and therefore potentially subject to regulation through the "significant new use rule" (SNUR) under the Toxic Substances Control Act (TSCA). This move means that companies and other entities that are currently manufacturing or importing nanomaterials must evaluate the implications of their activities under TSCA or risk enforcement action from the U.S.EPA. The U.S.EPA has also requested public comment on a petition filed to regulate nanoscale silver as a pesticide under the Federal Insecticide, Fungicide and Rodenticide Act.

8.2 Framework for Managing Risk in Occupational Environments

The traditional industrial hygiene paradigm of anticipation, recognition, evaluation, and control of hazards is used commonly to manage risks in occupational settings (NIOSH 2009). Depicted in Figure 8.3, anticipation and recognition involve the understanding of where and when contaminants may be generated, how they disperse into a workplace, and who may be exposed to them. Evaluation consists of measuring worker exposures or indicators of exposures, such as biological exposure indices. These measurements of exposure are then compared to working OELs that are based on toxicological and epidemiological data. Over the last fifty years, OELs have been established for many gases and particulate hazards that include airborne limits for inhalational hazards and biological exposure limits for all routes of exposure.

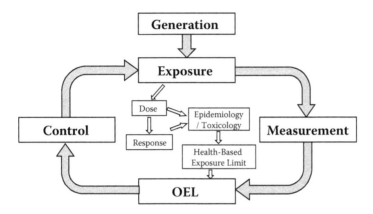

FIGURE 8.3
A classical industrial hygiene framework for managing risks in the occupational environment.

There are challenges in applying this traditional paradigm to managing the risks associated with working with engineered nanomaterials. The foremost challenge is the diversity of nanomaterials and the pace of their development. Regulatory OELs take many years to develop for a single compound, after adequate health data become available. As a result, only two RELs specific for engineered nanomaterials (titanium dioxide and carbon nanotubes, refer to Section 8.1.4) have been developed despite the fact that thousands of nanomaterials are used in the workplace. The EHS professional is, therefore, responsible to manage risk in an environment where no regulations exist and, more often than not, toxicity and epidemiological data are unavailable. Here, a review is provided on what is known regarding anticipation, recognition, evaluation, and control of engineered nanomaterials in the workplace. The discussion then turns to how these activities central to the field of industrial hygiene can be applied within various overall risk frameworks to manage risk.

8.2.1 Anticipation and Recognition

Brouwer (2010) and Kuhlbusch et al. (2011) summarize workplace processes that may lead to airborne exposures in the workplace. Production, handling, and manipulation of engineered nanomaterials define the rate of generation and dispersal into the workplace. Nanomaterials may be produced through mechanical processing or chemical processing, and etching can be used to break up or impart nanoscale features to bulk materials. Nano-sized particles may be grown by precipitation in liquid solution or by nucleation and condensation in gas-phase synthesis. Generation of particles into the air may result from handling feedstock or bursting bubbles or splashing of liquids. Typically, generation rates are low for production processes that are carried out in closed systems unless there are leaks in the system.

Generation rates may be considerably greater when handling and refining a nanomaterial after production. Relevant processes include harvesting, packaging, transfer, and equipment cleaning operations. Nanomaterial products must also be bagged and shipped. Bagging of dry powder nanomaterials has been observed as a source of airborne nanomaterials. Other processing steps include mixing powders in liquids or other matrices, such as epoxy, which can provide opportunities for generation of airborne particles.

Manipulation of larger objects with embedded or fixed nanomaterials can also result in generation and dispersal of these nanomaterials into the workplace. Manufacturing of products often requires that parts be sanded, cut, or manipulated in other ways. These processes may release the nanomaterials themselves or larger particles with the nanomaterial attached to their surfaces (Figure 8.2). Such processes are also required at the end of a product's life, such as in recycling.

As depicted in Figure 8.4, research laboratories, development facilities, production facilities, and waste-handling/recycling centers have potential for occupational exposures to engineered nanomaterials. Typically, a small number of researchers and/or technicians perform numerous processes with small quantities of potentially unique materials in the research laboratory. Similarly, staff in development facilities work with greater quantities of less compositionally variable materials in pilot-scale experiments to identify the processes, procedures, and products that will translate the novel discoveries made in the research laboratory into practical commercial end points. In production facilities, numerous workers may produce and handle large quantities of nanomaterials of a specific type. Workers in waste-handling and recycling centers are generally the end point for products that contain nanomaterials. Workers may interact with nanomaterials in a variety of handling procedures, such as fluid removal, shredding, grinding, and incineration. Each process offers opportunities for release of and exposure to nanomaterials.

8.2.2 Evaluation

8.2.2.1 Instrumentation

Evaluation of a workplace hazard involves assessing worker exposures. New methods to evaluate exposures and consensus strategies are an active area of research, as described by Ramachandran et al. (2011). For inhalational exposures, particle exposures are expressed as airborne concentrations (quantity of a contaminant divided by volume of air sampled,

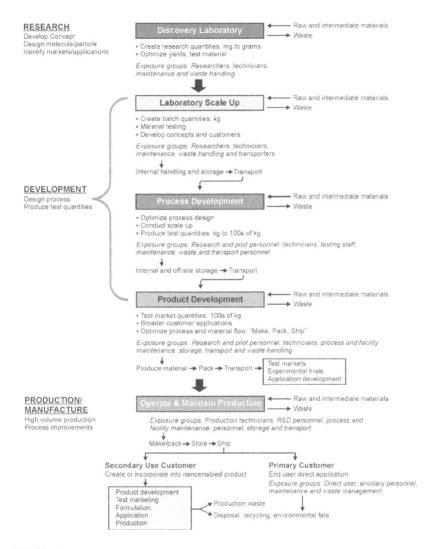

FIGURE 8.4
Life cycle of engineered nanomaterials in different types of workplaces with potential for occupational exposure from the research laboratory to manufacturing. (Reprinted from Schulte et al. [2008] with permission of Taylor & Francis.)

usually in m^{-3}). Airborne concentrations may be measured within the breathing zone of a worker, called a *personal sample*, or in the vicinity of a worker, called an *area sample*. Biological exposure indices, such as biomarkers in urine or blood, may be used to represent all routes of exposure, but these are not well developed for engineered nanomaterials.

Current methods to evaluate workplace concentrations of engineered nanomaterials include time-integrated samplers and direct-reading instruments. The NIOSH current intelligence bulletins (CIBs) for CNTs and TiO$_2$ specify collection of time-integrated samples with traditional size-selective personal samplers. In personal sampling, a belt-mounted sampling pump connected by a tube to a lapel-mounted sampler is used to aspirate air from within a worker's breathing zone. In the sampler, the air is passed through a filter that collects particles. The mass of all particles can be determined through gravimetric analysis, or a chemical-specific fraction of particles can be determined through various chemical analysis techniques.

Analysis of samples by mass alone is likely to be inadequate to characterize exposures for engineered nanomaterials because the nanomaterial may be obscured by other airborne particles present in the workplace atmosphere. In the CIB for TiO$_2$, NIOSH recommends a combination of gravimetric, chemical specific, and microscopic methods to assess exposures (NIOSH 2011). They recommend that two respirable samples be collected, with one analyzed for mass concentration through gravimetric sampling. No further actions are required if the mass concentration of all particles determined gravimetrically is less than the REL for ultrafine TiO$_2$ particles (0.3 mg m^{-3}). If the concentration is greater than this REL, further characterization is needed to determine the fraction of the respirable sample that is TiO$_2$ by bulk chemical analysis and its size distribution by transmission electron microscopy with energy dispersive spectroscopy (TEM-EDS).

For carbon nanotubes and nanofibers, NIOSH recommends analysis of collected particles by Method 5040 to determine the mass of elemental carbon, because that is the primary component of these nanomaterials (NIOSH 2010). However, there are other sources of elemental carbon in the workplace, such as diesel exhaust or other combustion sources. Thus, NIOSH recommends that transmission electron microscopy (TEM) with energy dispersive spectroscopy (EDS) be used to verify that the elemental carbon present in a sample is indeed nanomaterial.

These recommended evaluation methods highlight the need to differentiate between engineered nanomaterials and other particles that occur incidentally in the workplace. However, they are complicated by a reliance on traditional sampling (i.e., respirable samplers) and analysis techniques. Electron microscopy is an effective tool to characterize airborne particles by size, morphology, and composition, although it is expensive and time consuming for routine measurements and subject to analytical error (Powers et al. 2007).

New sampling methods that may streamline sample collection are being developed. For example, a new personal sampler described by Cena, Anthony, and Peters (2011) samples only those nanoparticles apart from larger respirable particles in a way that mimics their behavior in the human respiratory tract. By capturing only nanoparticles on the sampling media, cost-efficient bulk analysis techniques (e.g., inductively coupled plasma mass spectrometry) can be used to estimate the amount of deposited nanoparticles without the need for electron microscopy.

The limitations of respirable samplers have made direct-reading instruments more attractive for evaluating exposures to engineered nanomaterials. Table 8.1 shows images of some direct-reading instruments used in monitoring of engineered nanomaterials in the workplace. Two instruments in particular have been consistently recommended for this purpose: the condensation particle counter (CPC), which measures particles from about 10 nm to 1000 nm; and the optical particle counter (OPC), which measures particles from 300 nm to roughly 10 µm (Table 8.1A). These instruments can be used to estimate ultrafine particle number and respirable mass concentrations (Peters et al. 2006; NIOSH 2009). This information is valuable because it enables large and small particles to be "seen" by the EHS professional and can be used to identify areas where controls are needed and to evaluate their effectiveness once installed. They are, however, limited to area measurements because of their size.

Smaller direct-reading instruments have recently become available for measuring personal exposures (Table 8.1B). A new class of instruments based on electrical measurement of charged particles with low-cost upstream size classification allows personal measurement of nanoparticle-number concentration and geometric mean diameter. An example of this type of monitor is described by Fierz et al. (2010) and has recently been commercialized as the DiSCmini from Matter Engineering. This instrument, paired with a personal photometer, can be used to simultaneously measure personal exposures in metrics of number and mass concentration. Such measurements are valuable in identifying the determinants of personal exposure for a wide range of particulate contaminants in the workplace.

8.2.2.2 Monitoring Strategies

A tiered approach with screening, extensive characterization, and routine monitoring is useful in characterizing a workplace where engineered nanomaterials are present (NIOSH 2009; Ramachandran et al. 2011). A CPC and an OPC can be used with aerosol-mapping monitoring to screen for areas of high concentration within a workplace (Peters et al. 2006). Alternatively, these instruments can be used with task-based monitoring to determine specific tasks associated with elevated exposures (Methner, Hodson, and Geraci 2010; Vosburgh et al. 2011).

Extensive characterization can be conducted to further evaluate airborne particles in areas of concern identified through screening. Particle samples can be collected and analyzed chemically or by microscopy to determine the extent to which concentrations measured with the

TABLE 8.1

Some Battery-Powered, Direct-Reading Instruments That Have Been Used to Assess Exposures in the Workplace

Type of Instrument	Specifications	Example
A: Area Monitors		
Condensation particle counter (CPC)	Outputs number concentration of particles from 10 nm to ≈1000 nm	TSI Handheld CPC, Model 3007
Optical particle counter (OPC)	Number concentration of particles from 300 nm to ≈10 μm	Grimm PDM 1.108
B: Personal Monitors		
Diffusion charger with upstream classifier	Number concentration and median number diameter for particles smaller than 300 nm	Matter Engineering DiSCmini
Photometer	Mass concentration of particles from 500 nm to ≈10 μm	TSI AM510 Personal Aerosol Monitor

direct-reading instruments are actually engineered nanomaterial. Other more expensive, size-resolved particle instruments (e.g., scanning mobility particle sizer or aerodynamic particle sizer) may also be used to determine the size distribution of the aerosol in the workplace.

Routine monitoring methods can be devised after extensive characterization is carried out. Routine monitoring is performed to ensure that airborne concentrations stay below acceptable levels established by the facility. Routine monitoring involves collection of specific data, either from real-time monitors or time-integrated samplers, which allows the workplace concentration of engineered nanomaterials to be established. This type of monitoring can be used to track worker exposure or control effectiveness over time.

8.2.3 Control

As shown in Figure 8.5, a hierarchical approach is recommended when applying controls to reduce concentrations to reduce risk (Schulte et al. 2008). A variety of control strategies are implemented in workplaces that produce or handle engineered nanomaterials, as described by Dahm, Yencken, and Schubauer-Berigan (2011) for the carbonaceous nanomaterial industry. Elimination or substitution of the engineered nanomaterial should be considered first. A nanomaterial may be coated, or encapsulated, without hampering performance to reduce biological activity and the likelihood of emission into a workplace. If these control solutions are impractical, then processes should be isolated to minimize the number of workers who come into contact with the nanomaterial. For example, a process may be placed in an enclosure with physical barriers, such as a glove box, or in a separate room, where only those workers required to tend to the process are allowed.

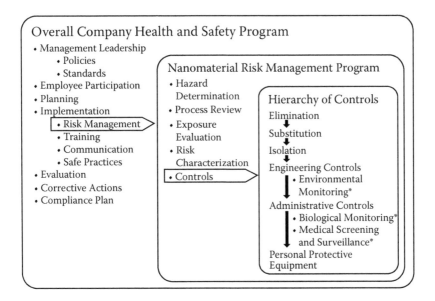

FIGURE 8.5
Hierarchy of controls nested within an occupational safety and health management system designed to minimize worker exposure to contaminants. (Reprinted from Schulte et al. [2008] with permission of Taylor & Francis.)

Engineering controls, such as the use of local exhaust ventilation, are next on the hierarchy of controls. Generally, these controls are most effective when applied close to the emission source, serving to interrupt the path of a contaminant from the source to the breathing zone. Nanomaterials follow well-known aerosol physics behaviors. The effectiveness of various hoods for capturing particles is well documented in ventilation manuals. However, interest in the control of nanomaterials in the workplace has prompted new research, which has shown (a) that certain hoods are more effective than others for nanomaterials and (b) that the way that they are configured alters their capture efficiency (Tsai, Huang, and Ellenbecker 2010).

Personal protective equipment (PPE), such as the use of face-filtering respirators, should be considered as a last choice of controls because it places the responsibility for preventing illness on the worker (Schulte et al. 2008). Moreover, PPE is often uncomfortable, interferes with communication and vision, and is effective only when properly used and requires implementation of a well-managed, complete, and systematic program as per OSHA guidelines. Most PPE exhibits a characteristic dip where diffusion and impaction are least effective at causing a particle to deviate and hit a fiber—called the most penetrating particle size. This behavior is dependent largely upon particle size rather than composition. Thus, the effectiveness of respirators for nanomaterials generally follow experimental results obtained for other compounds (Rengasamy, Miller, and Eimer 2011).

8.2.4 Overall Risk Management

Risk management involves coherent synthesis of anticipation, recognition, evaluation, and control. There are, however, considerable uncertainties in developing risk management strategies for engineered nanomaterials. Few occupational exposure limits have been established owing to the limited toxicological and epidemiological data and the ever-growing list of nanomaterials in the workplace. Sampling methods are imperfect for assessing exposures, and few measurements have been made in workplaces where nanomaterials are handled.

Despite these uncertainties, management strategies have been proposed for managing the risks of engineered nanomaterials in the workplace (NIOSH 2009; Groso et al. 2010). NIOSH's *Approaches to Safe Nanotechnology* provides a monitoring-based strategy that describes how the basic principles of industrial hygiene, presented in Section 8.2, can be applied in a cohesive way to manage risk. The document outlines potential health concerns and exposure routes, safety hazards, ways to assess exposures in the workplace, guidelines for working with nanomaterials, and methods for occupational health surveillance.

A hazards-based approach may be used to implement precautionary measures consistent with good EHS practices. Control banding allows a rational decision-making process for prioritizing and implementing controls in the absence of complete information (Zalk, Paik, and Swuste 2009). Control banding combines information about a particular nanomaterial (e.g., surface chemistry, shape, diameter, solubility, carcinogenicity, and toxicity) with the likelihood that an exposure will occur. Efforts can then be focused first on the most severe exposure situations that would result in serious adverse health effects and are likely to occur.

8.3 Environmental Health and Safety Needs

The large number of reports generated during the past few years by national and international governmental agencies and research organizations on the issue of EHS needs for nanomaterials reflects both the great importance and the information gaps regarding this issue. The following sections provide brief summaries of some recent reports.

8.3.1 NIOSH Critical Topic Areas

NIOSH has defined ten critical topic areas to address gaps in knowledge, risk management strategies, and the development of recommendations for engineered nanomaterials in the workplace (NIOSH 2009). These topics include the need for research in all areas of traditional industrial hygiene, including anticipation, recognition, evaluation, and control. For anticipation and recognition, NIOSH defines that workplace exposure assessments are needed to identify determinants of exposure for inhalation and dermal routes. For evaluation, they define a critical need to evaluate and develop methods to accurately measure nanomaterials in the workplace. For control, critical topics include evaluation of engineering controls and PPE for effectiveness. Other critical topic areas include development of information on epidemiology and surveillance, toxicity and internal dose, risk assessment, and fire and explosion safety. They also stress the importance of translating this information into recommendations and guidance.

8.3.2 National Nanotechnology Initiative Environmental Health and Safety Research Needs for Engineered Nanoscale Materials

The National Nanotechnology Initiative (NNI), first formed in the mid-1990s from informal meetings among staff members from several agencies, is a federal government research and development program established to coordinate the efforts of twenty-six participating federal agencies regarding nanoscale science, engineering, and technology. The National Science Engineering and Technology Subcommittee of the National Science and Technology Council identifies the research and information needed to address environmental, health, and safety issues regarding engineered nanomaterials (NNI 2011).

This strategy broadly applies to protecting health in the environment and workplace. The section on risk management methods provides a good summary of the current research and information needs in several important areas:

- Improved understanding of the challenges that airborne nanomaterials present for process design and engineering control systems
- Understanding and development of manufacturing approaches that minimize environmental impact to enable green design principles
- Determination of the stages in a product's life cycle that introduce the potential for EHS risks
- Evaluation of whether current risk communication methods are adequate for known risks and for risks that can be anticipated from currently available information

8.3.3 U.S. Environmental Protection Agency White Paper on Nanotechnology

Under its federal mandate, the U.S. Environmental Protection Agency (U.S.EPA) has the regulatory authority to protect the environment from a variety of possible hazards, such as those that may be presented by nanomaterials. Under the Toxic Substances Control Act (TSCA), U.S.EPA can require premanufacture notification (PMN) from manufacturing companies for products containing unregistered substances. TSCA provides the U.S.EPA with the authority to identify and control "new chemicals" that may pose a threat to human health or the environment, and the U.S.EPA can determine which nanomaterials meet this criterion. As an example of its regulatory authority under the Federal Insecticide, Fungicide and Rodenticide Act (FIFRA), in late 2006 the U.S.EPA required review of a Samsung clothes washer (discussed in Chapter 3) based on the manufacturer's claim to add nanoscale silver as an antibacterial agent. Other U.S.EPA regulatory authority through the Comprehensive Environmental Response, Compensation, and Liability Act (CERCLA), Resource Conservation and Recovery Act (RCRA), and the Clean Air Act (CAA) may also impact the manufacture of products containing nanomaterials.

The U.S.EPA recently released its final version of a white paper on nanotechnology (2007). This report focuses primarily on the potential environmental impacts of nanomaterials, but it has implications for companies that manufacture and use nanomaterials as part of their operations and may release nanomaterials into the environment.

8.3.4 Voluntary Standards

As discussed in Chapter 9, several organizations are developing voluntary standards for nanomaterials, and many of these include EHS standards. The American Society for Testing and Materials International (ASTM International) formed Committee E56 on Nanotechnology in October 2005 in response to the recognition by the organization that it needed to specifically address this emerging technology area. A recent work product of Committee E56 is a terminology standard related to nanotechnology (ASTM International 2006). This document is a critical first step to define nanotechnology and the related terms that are used to describe the materials and their characteristics. The E56 subcommittee on EHS, whose focus would include worker safety issues, has not yet provided any published documents on this area. The ISO Technical Committee 229 (TC229) for Nanotechnologies working groups include: terminology and nomenclature; measurement and characterization; and health, safety, and environmental aspects of nanotechnologies. The working groups are currently developing technical reports that will contribute toward production of a standards document for nanotechnologies by the TC229. Those sections that deal with the EHS aspects of nanotechnologies will be most relevant to the concerns addressed in this chapter regarding worker exposures to nanomaterials.

8.4 Ongoing Governmental Efforts on Environmental Health and Safety

In addition to these recent reports, other EHS efforts are ongoing in the United States and globally.

8.4.1 Occupational Safety and Health Administration

The U.S. Occupational Safety and Health Administration (OSHA), whose mission is to assure the safety and health of America's workers by setting and enforcing standards, has not yet developed guidance documentation or specific standards for nanotechnology and nanomaterials. However, OSHA does participate in the NNI. OSHA plans to develop guidance for employers and employees engaged in operations involving nanomaterials. A clear drawback of the traditional regulatory and standards development processes by OSHA is the long time frame and the amount of research and data required to develop standards.

In the absence of published standards or guidance documents specifically for nanotechnology, some elements of current OSHA regulations are now applicable for nanomaterials. Worker training for use and handling of nanomaterials is needed to ensure the compliance under the right-to-know requirements. In addition, employers must comply with the general duty clause that requires them to provide each employee a place of employment that is free from recognized hazards that are causing or are likely to cause death or serious physical harm (OSHA Act 29U56 654 Section 5(a)(1)). With regard to respiratory protection, OSHA has a standard (OSHA 24 CFR 1910, 134) that could be applicable to nanomaterials, but the small size of nanomaterials may present a challenge to the effectiveness of traditional personal protective equipment designed to protect workers. As previously discussed, the adequacy of existing respiratory protective equipment and its filtration materials for protecting against nanomaterials exposures is an area of active research.

8.4.2 NIOSH Efforts

NIOSH has devoted a considerable portion of its internal resources to ensuring the protection of those who work with engineered nanomaterials. As presented throughout this chapter, NIOSH's activities include all aspects of managing risks in the workplace, from development of RELs (Section 8.1.4) to guidance for managing risk in workplaces (Section 8.2.4). NIOSH has devoted teams of researchers to work on various aspects of each of the ten critical areas (see Section 8.3.1) that they identified in their "Approaches to Safe Nanotechnology" document (NIOSH 2009). Their field team may be of particular interest to the readers of this book. As described at http://www.cdc.gov/niosh/docs/2008-120/, the field team offers their expertise, services, instrumentation, and unbiased assessments at no cost to companies. This service has helped companies develop and implement management strategies for engineered nanomaterials in the workplace. The information gained through the activities of the field team are used to revise recommendation and guidance documents.

8.4.3 The European Union and Registration, Evaluation, and Authorization of Chemicals (REACH)

In Europe, one area of interest is how nanoparticles are likely to be treated under the European Union's Registration, Evaluation, and Authorization of Chemicals (REACH) program. Because nanomaterials fall within the scope of REACH, their health and environmental properties must be assessed in accordance with the provisions of this regulation. However, the European Union (EU) countries are in consensus with the international recognition that methodologies for identifying hazards and evaluating risks of substances at the nanoscale need to be further refined over the next few years. The European Commission is funding research projects to assess the health and environmental impacts of nanoparticles under the Seventh Research Framework Program. It will also be necessary to carefully monitor over the next few years whether the threshold for registration and the information requirements under REACH are adequate to address potential risks from engineered nanoparticles.

Other European organizations have also addressed the issue of worker safety and nanotechnology as part of their recent reports (Organization for Economic and Commercial Development 2006; Scientific Committee on Emerging and Newly Identified Health Risks 2006). Further discussion of the findings and recommendations from these reports can be found in Chapter 9.

8.5 Summary

Nanotechnology promises novel materials with innovative properties designed to enhance a diverse array of commercial products. The introduction of nanomaterials into the workplace presents considerable uncertainty for management of risk. Workers are on the front line of exposures to these novel and unique materials. It is imperative that appropriate actions and funding be directed toward obtaining the information needed to develop relevant guidance to protect their health and safety.

A proactive approach is being taken, both nationally and internationally, to address worker safety issues in the nanotechnology environment. Governmental agencies have identified research gaps required to reduce uncertainties and more effectively manage the risks of engineered nanomaterials in the workplace. Extensive research is being conducted to fill these gaps. These proactive strategies are laudable and will result in mitigation of risk. However, the challenge is great, owing to the complexity of nanomaterials and the rapid pace of their development.

References

Abe, S. 2011. Towards a standard vocabulary for nanomaterials relative to regulatory science. *Genes and Environment* 33(1): 10–13.

Brouwer, D. 2010. Exposure to manufactured nanoparticles in different workplaces. *Toxicology* 269(2–3): 120–27.

Cena, L. G., R. Anthony, and T. M. Peters. 2011. A personal nanoparticle respiratory deposition (NRD) sampler. *Environ. Sci. Technol.* 45: 6483–90.

Cena, L. G., and T. M. Peters. 2011. Characterization and control of airborne particles emitted during production of epoxy/carbon nanotube nanocomposites. *J. Occup. Environ. Hyg.* 8(2): 86–92.

Dahm, M. M., M. S. Yencken, and M. K. Schubauer-Berigan. 2011. Exposure control strategies in the carbonaceous nanomaterial industry. *J. Occup. Environ. Med.* 53: S68.

Fierz, M., C. Houle, P. Steigmeier, and H. Burtscher. 2010. Design, calibration, and field performance of a miniature diffusion size classifier. *Aerosol Sci. Technol.* 45(1): 1–10.

Groso, A., A. Petri-Fink, A. Magrez, M. Riediker, and T. Meyer. 2010. Management of nanomaterials safety in research environment. *Particle Fibre Toxicol.* 7(1): 40.

Kuhlbusch, T. A. J., C. Asbach, H. Fissan, D. Göhler, and M. Stintz. 2011. Nanoparticle exposure at nanotechnology workplaces: A review. *Particle Fibre Toxicol.* 8(1): 22.

Lux Research. 2006. *How industry leaders organize for nanotech innovation.* New York: Lux Research.

———. 2009. *Nanomaterials state of the market Q1 2009.* New York: Lux Research.

Maynard, A. D., D. B. Warheit, and M. A. Philbert. 2011. The new toxicology of sophisticated materials: Nanotoxicology and beyond. *Toxicological Sciences* 120(suppl. 1): S109.

Methner, M., L. Hodson, and C. Geraci. 2010. Nanoparticle emission assessment technique (neat) for the identification and measurement of potential inhalation exposure to engineered nanomaterials: Part A. *J. Occup. Environ. Hyg.* 7(3): 127–32.

NIOSH. 2009. Approaches to safe nanotechnology. http://www.cdc.gov/niosh/docs/2009-125/pdfs/2009-125.pdf

———. 2010. Draft current intelligence bulletin: Occupational exposure to carbon nanotubes and nanofibers. http://www.cdc.gov/niosh/docket/review/docket161A/pdfs/carbonNanotubeCIB_PublicReviewOfDraft.pdf

———. 2011. Current intelligence bulletin 63: Occupational exposure to titanium dioxide. http://www.cdc.gov/niosh/docs/2011-160/pdfs/2011-160.pdf

NNI. 2011. National Nanotechnology Initiative: Environmental, health, and safety research strategy. National Science and Technology Council Committee on Technology. Washington, DC.

NSET. 2010. The National Nanotechnology Initiative: Research and development leading to a revolution in technology and industry supplement to the president's FY2011 budget. Subcommittee on Nanoscale Science, National Science and Technology Council. Washington, DC.

Peters, T. M., W. A. Heitbrink, D. E. Evans, T. J. Slavin, and A. D. Maynard. 2006. The mapping of fine and ultrafine particle concentrations in an engine machining and assembly facility. *Annals Occup. Hyg.* 50(3): 249–57.

Powers, K. W., M. Palazuelos, B. M. Moudgil, and S. M. Roberts. 2007. Characterization of the size, shape, and state of dispersion of nanoparticles for toxicological studies. *Nanotoxicology* 1(1): 42–51.

Ramachandran, G., M. Ostraat, D. E. Evans, M. M. Methner, P. O'Shaughnessy, J. D'Arcy, C. L. Geraci, E. Stevenson, A. Maynard, and K. Rickabaugh. 2011. A strategy for assessing workplace exposures to nanomaterials. *J. Occup. Environ. Hyg.* 8(11): 673–85.

Rengasamy, S., A. Miller, and B. Eimer. 2011. Evaluation of the filtration performance of NIOSH-approved N95 filtering facepiece respirators by photometric and number-based test methods. *J. Occup. Environ. Hyg.* 8(1): 23–30.

Schulte, P., C. Geraci, R. Zumwalde, M. Hoover, and E. Kuempel. 2008. Occupational risk management of engineered nanoparticles. *J. Occup. Environ. Hyg.* 5(4): 239–49.

Tsai, S. J. C., R. F. Huang, and M. J. Ellenbecker. 2010. Airborne nanoparticle exposures while using constant-flow, constant-velocity, and air-curtain-isolated fume hoods. *Annals Occup. Hyg.* 54(1): 78.

Vosburgh, D. J. H., D. A. Boysen, J. J. Oleson, and T. M. Peters. 2011. Airborne nanoparticle concentrations in the manufacturing of polytetrafluoroethylene (PTFE) apparel. *J. Occup. Environ. Hyg.* 8(3): 139–46.

Zalk, D. M., S. Y. Paik, and P. Swuste. 2009. Evaluating the control banding nanotool: A qualitative risk assessment method for controlling nanoparticle exposures. *J. Nanoparticle Res.* 11(7): 1685–1704.

9

Nanotechnology Risk Communication

Ann Bostrom and Ragnar Löfstedt

CONTENTS

9.1 Introduction

As recently as 2010, friends and colleagues were skeptical that a mental models study of nanotechnology would be possible; they didn't think people had mental models of nanotechnology. This inspired one of our research groups to attempt such a study. In initial explorations of how to go about this, it seemed a good idea to examine how consumers react to nano-products. At the local consumers cooperative in Seattle, both sunscreen containing nano zinc oxide or nano titanium dioxide and socks fabricated to contain nanosilver as an antimicrobial were available for sale. Expeditions to various stores led the team to conclude that labeling was highly variable, with sunscreens sometimes labeled as containing micronized ingredients, or not indicating whether or not they were nano-products, regardless of content. It became obvious early on that for the team's research, we would have to order products online to obtain a sufficient quantity of a product clearly labeled as a nano-product, since we wanted study participants to be able to take home the product of their choice.

An inveterate label reader, the lead author found in subsequent forays to the co-op through the following year that sunscreen labels were evolving rapidly, sometimes specifying particle sizes above 100 nm—implying they weren't nanoparticles—though it seemed evident that the sunscreens were functionally nano-products (see Shatkin et al. 2010 for a discussion of functionality). The consumers cooperative also stopped selling the no-stink socks, though it still carried that brand of socks. In May 2009, the consumers cooperative newsletter had written the following about nanotechnology:

> Nanoparticles are nothing new. They've been around for millions of years. A nanoparticle is anything that measures 100 nm (nanometer) or less. A nanometer is one billionth of a meter. A human hair is 80,000 nm wide. Hemoglobin, a component of red blood cells, measures 5 nm.

The article included an inset textbox with the alternative view that "overwhelming evidence that these particles are safe simply is not there" (McCully 2009). The National Organic Standards Board and National Organic Program met in Seattle in April 2011, when they addressed four topics, including nanotechnology. About this, the consumers cooperative (PCC 2011) wrote a piece that opened with the sentiment and some of the exact same language that "nanoparticles are nothing new," but included the statement:

> If there's any hole at all in the Natural Products Association (NPA) Natural Standard (which PCC adheres to as our standard for supplements and body care products), it's that it does not require a company to reveal the presence of nanoparticles. (PCC 2011)

This experience illustrates consumer lack of awareness (the research team's), uninformative labeling, explicit framing of nanotechnology by the media, shifting advocacy positions, and the absence of standards regarding nanotechnology risks. These characterize current nanotechnology risk communication, and are explored in more detail in this chapter. The chapter opens with a description of the bigger decision and information context, then explores labeling issues, delves into how people make judgments about nanotechnology in the absence of specific information, provides a snapshot of recent research on awareness and decisions about nanotechnology, and concludes with an appeal for more risk communication about nanotechnology and more science in nanotechnology risk communication.

9.2 Decision Making: Follow the Money

It appears that decisions about nanotechnology are being made based on money, not on risk. A recent primer on nanotechnology has this to say about nanotechnology market prospects and media reports:

> The U.S. National Science Foundation has estimated that the global nanotechnology market could be worth US$1 trillion by 2015. In parallel, much has been written and presented about the excitement and possible dangers of these materials. The tone of these media articles range from how these wonder materials are going to revolutionize all aspects of our lives to how they might kill us! (Salomon, Courtney, and Shuttler 2010, 1)

Contrary to the benefit–risk balance implied in this quote, reports on potential risks have in fact barely tarnished the glow on the widespread technological enthusiasm for nanotechnology. Media analyses show that risk actually constitutes only a small proportion—in the single-digit percentages—of newspaper articles on nanotechnology in the last decade (e.g., Bostrom and Löfstedt 2010; Friedman and Egolf 2011). Perhaps even more noteworthy, governmental initiatives on nanotechnology dedicate only a small fraction of their funds to environmental health and safety (Bostrom and Löfstedt 2010). While the United States has invested over $14 billion in its National Nanotechnology Initiative since its inception in 2001, of that, only a small proportion—albeit increasing—has been dedicated to researching environmental, health, and safety aspects of nanotechnology (US$37.7 million in fiscal year 2006, $123.5 million in fiscal year 2012) (Holdren, Sunstein, and Siddiqui 2012; Sargent 2011).

In sum, despite occasional calls for a moratorium on some nanotechnology applications (e.g., in food, FOE 2008), and even legal battles (see Shatkin, Chapter 5 of this book, for an example; cf. Devries and Jehl 2009), risk communication has played a peripheral role in nanotechnology communications to date.

9.3 Informing Risk Decisions: Labeling

As John F. Kennedy proclaimed in 1962, consumers have a right to safety, to be informed, to choose, and to be heard (Throne-Holst and Strandbakken 2009; Kennedy 1962). Risk communication—including specific forms of communications ranging from labeling to stakeholder engagement—is called for by nanotechnology regulatory strategies in some nations (Holdren et al. 2011). Transparency is a hallmark initiative of the Obama administration, and access to information, transparency, and meaningfulness to nonscientists are emphasized in the European Commission Code of conduct for responsible nanoscience and nanotechnologies research. Implementation of these good intentions, however, is on a bumpy road.

Labeling requirements have met obstacles, and existing labeling appears to be either deficient or disappearing. A move in the European Parliament to require food containing nano-ingredients to be labeled for consumer information stalled when an agreement could not be reached in 2011. And an Australian analysis of material safety data sheets (MSDS) and labels pertaining to engineered nanomaterials found that "most of the MSDS evaluated (84%) did not provide adequate and accurate information sufficient to inform an occupational risk assessment for nanomaterial contained in the product" (Safe Work Australia 2010). A 2007 study found nanoparticles in sunscreens that were not labeled nano (Consumers Union 2008),* and there is suggestive evidence that cosmetics are less likely now to be labeled as containing nanoparticles than they were a few years ago in the United States (Throne-Holst and Strandbakken 2009). It is tempting, if cynical, to conclude that proprietary marketing research may have reached the same conclusion as recent experimental research: Nano-labeling of sunscreen products may increase perceptions of risk (Siegrist and Keller 2011). Nanotechnology scholar Dietram Scheufele remarked in his nanopublic blog that manufacturers were removing nano labels from product descriptions in anticipation of regulatory action from the U.S. Environmental Protection Agency (EPA) (Scheufele 2006).

Product labels are immediately relevant to purchase-and-use choices, and are a familiar and globally accessible path to reach consumers with risk information. With the advent of smart phones that read quick-response (QR) codes, labels can provide access to virtually unlimited information about products. To date, however, nano-product labels have not even begun to fulfill this promise.

To wit, when a 2011 Australian government survey asked: "If you have heard or read stories about possible risks of using sunscreens with nanoparticles in them, how well do you feel you understand the risk issues?" only 9% of respondents reported completely, 21% mostly, 27% somewhat, and the remaining 43% responded with "a little" or "not at all" ($n = 1,000$; Australian Govt. 2012).

---

* Nanotechnologists suggested at the September 2008 SRA nano risk workshop (Shatkin and North 2010) that increased demand for nanoparticles may have shifted market shares to such an extent that non-nanoparticles of, for example, titanium dioxide, are increasingly difficult to obtain for products like sunscreen. Consumer Reports commissioned tests of five brands of sunscreen that company representatives stated did not contain nanosize particles of titanium dioxide or zinc oxide. Four of them, all labeled natural or organic, actually did contain nanoparticles (Consumers Union 2008).

9.4 Imagining Risk

Nanotechnology and its potential risks are impossible to perceive without descriptions and visualizations—that is, communications. Nanotechnology is intangible, invisible, scientifically complex, and replete with risk unknowns; it promises solutions to many of the world's ills and has a wildly touted economic potential (Bostrom and Löfstedt 2010; Sargent 2011). For these reasons, artists and scientists have invested in visualizing nanotechnologies (Ruivenkamp and Rip 2010), and commercial (Salamon et al. 2010), advocacy, and nonprofit (e.g., Maynard 2007, 2009a,b) and governmental organizations (e.g., Sargent 2011) have tried to simplify and translate nanotechnology for consumers and citizens, resulting in a plethora of primers.

In one example of these primers, the ETC Group (action group on Erosion, Technology and Concentration) issued "A Tiny Primer on Nano-Scale Technologies and the Little BANG Theory" in 2005 (ETC 2005). A Tiny Primer focuses on issues of social and economic control:

> With only a reduction in size and no change in substance, materials can exhibit new properties such as electrical conductivity, elasticity, greater strength, different colour and greater reactivity.... While "Grey Goo" has grabbed the headlines in the media (where self replicating nano-scale mechanical robots escape control until they wreak havoc on the global ecosystem), the more likely future threat is that the merger of living and non-living matter will result in hybrid organisms and products that are not easy to control and behave in unpredictable ways.

While the effects of these primers have apparently not been assessed specifically, many of them perpetuate common frames and mental models of nanotechnology, as illustrated by gray goo in the preceding excerpt.

John Sladek's *The Reproductive System* (aka *Mechasm*) in 1968 foreshadowed the infamous concept of gray goo:

> Wompler's Walking Babies aren't selling like they used to, so the company develops Project 32, producing self-replicating mechanisms designed to repair inter-cellular breakdowns. But then the metal boxes begin crawling about the laboratory, feeding voraciously on metal and multiplying.

The continued popularity of this idea decades later illustrates how a simple mental model can catch on (for another example of this, see Drexler 1986; Osborne 2003; cf. Friedman and Egolf 2011).

Simple mental models, metaphors, and analogies are the building blocks of science as well as of public inference and imagination (Bostrom 2008; Bostrom et al. 1992; Gentner and Stevens 1983; Gentner, Holyoak, and Kokinov 2001; Morgan et al. 1992, 2002; Nersessian and Chandrasekharan 2009). Analogies build on deep structural similarities between objects or events and can inspire new insights and understanding. In their in vivo studies of science in the lab, Dunbar and Blanchette (2001) demonstrate that scientists' use of types of analogies changes with their goals. So such analogies and metaphors are significant, not least because they are often inherently appealing—easy to visualize. Scientists have, for example, adopted an asbestos metaphor for carbon nanotubes (CNTs), as illustrated by this journal article title: "Carbon Nanotubes Introduced into the Abdominal Cavity Display Asbestos-Like Pathogenic Behaviour in a Pilot Study" (Poland et al. 2008).

The asbestos analogy was adopted by many journalists; in 2000–2004 in the United States, 13% of newspaper reports on nanotechnology risks compared them to asbestos, and in the United Kingdom 16% (Friedman and Egolf 2005). Figures 9.1 and 9.2 show how the presence of nanotechnology in the media has grown since the turn of the century as well as the parallel growth in prevalence of the asbestos analogy, at least until 2008. Exemplifying the kinds of articles produced by these searches, a May 20, 2008, report in the *Guardian Unlimited* (England) included the statement that "carbon nanotubes may be as hazardous to health as asbestos," and a physical comparison between carbon nanotubes and asbestos fibers is in its description of an article from *Nature Nanotechnology*. However, the overall proportion of all nanotechnology (not necessarily on risk) articles using the analogy was minuscule, far less than 1% (see Figures 9.1 and 9.2).

Inducing a frame of reference, for example by using a dominating comparison like the comparison with asbestos, is nevertheless a potentially potent way of focusing attention (Nisbet and Scheufele 2007). Researchers and governments have convened citizen juries, consensus conferences, and public focus groups on nanotechnology around the world, for example in Australia and Germany (Bowman and Hodge 2007), New Zealand (Cook and Fairweather 2005), Switzerland (Bellucci and Burri 2008; TA-Swiss 2006), the United Kingdom (Pidgeon and Rogers-Hayden 2007; Pidgeon et al. 2009), and the United States (Hamlet, Cobb, and Guston 2008; Kleinman and Powell 2005; Pidgeon et al. 2009). Personal experiences and analogies with other technologies, such as genetic modification and vaccines, and with other risks such as asbestos, were salient in some of these nanotechnology discussions and framed the consideration of social and ethical issues (Cook and Fairweather 2005; TA-Swiss 2006).

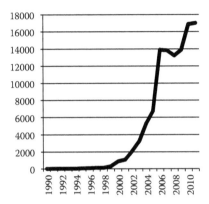

ACCESS World News Media database Nov 2011 search results for (nanoparticle* or nanomaterial* or nanotech* or nanoscience*) in all text not iPod

FIGURE 9.1
News media analysis November 2011. Search of ACCESS World news media with "(nanoparticle* or nanomaterial* or nanotech* or nanoscience*)."

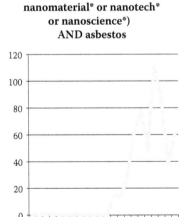

(nanoparticle* or nanomaterial* or nanotech* or nanoscience*) AND asbestos

FIGURE 9.2
News media analysis November 2011. Search of ACCESS World news media with "(nanoparticle* or nanomaterial* or nanotech* or nanoscience*) AND asbestos."

9.5 Prevalence of Products, Absence of Evidence, and Awareness

In a 2005 market research report, nanotechnology-based delivery systems were portrayed as the prime area of innovation in the U.S. skin-care market over the past decade. Lux Research projected in the same report that by 2014, nanotechnology-based products will be a $2.6-trillion business (Packaged Facts 2005). Reviews show that nanoparticles are commonly found in sunscreens (EWG 2009), with reports of over 300 nanoparticle sunscreen products on the market in 2006 (Garber and Berger 2006). In Australia, which has the highest skin cancer rates in the world, over 250 registered sunscreens contained nanoparticles of titanium dioxide in 2006 (Australian Govt. 2006); in 2011 it was estimated that a third of the 1,000 sunscreens on the market in Australia contain nanoparticles (Harvey, cited in Australian Govt. 2011a). Consensus has yet to emerge on the risk implications of this.

The U.S. advocacy organization Environmental Working Group (EWG) judged consumer exposures to nanoparticles in sunscreens to be of less concern than those stemming from toxicological properties of other sunscreen ingredients (EWG 2009). They concluded, based on existing peer-reviewed studies, that nanoscale titanium dioxide sunscreen is unlikely to be absorbed through human skin, though ingestion and inhalation may be of concern. EWG (2008) also noted that the United States lags behind Europe in screening and approving sunscreen technologies.[*] However, gaps in inhalation exposure data and toxicology led some researchers to deem currently possible nano-TiO_2 (nTiO_2) risk assessments inadequate for regulatory purposes (Christensen et al. 2011; see also Shatkin et al. 2010). In August 2006, the EPA found that the titanium nanoparticles used in sunscreens caused neurological changes in mice; a more recent study found DNA damage (Trouiller et al. 2009), although yet another study found no DNA damage to human blood lymphocytes exposed to nTiO_2 (Hackenberg et al. 2011). Because exposure to nTiO_2 from applying sunscreen may not result in skin penetration, and other exposure routes may be less likely or less important, it may not harm human health (Australian Govt. 2006; Bennat et al. 2000; Butz et al. 2007; Maynard et al. 2006; Schulz et al. 2002).

---

[*] In August 2008, subsequent to the release of the EWG review, Senators Dodd of Connecticut and Reed of Rhode Island introduced the Sunscreen Labeling Protection Act of 2008 to address the lack of sunscreen safety standards in the United States (Singer 2008). While this was an attempt to force the FDA to make final rules that the agency had proposed and did not directly pertain to nanoparticle ingredients, it highlights the demand for risk-related information on sunscreens.

It remains unclear whether or not nanoparticles are able to penetrate lesions, for example, those created by sunburn (Clift 2008). Ecological risks are also unknown. Nevertheless, nonprofits and advocacy groups have petitioned the U.S. Food and Drug Administration (FDA) to regulate sunscreens more stringently, with some demanding the recall of nanotechnology products, and others demanding that the FDA require full safety assessments of the use of engineered nanomaterials.

In a nationally representative survey of 1,000 adults in the United States in April 2009, 69% reported using sunscreen at least occasionally, and 31% reported wanting a clear, see-through sunscreen (Consumers Union 2008). Nano zinc oxide and nano titanium dioxide have this property. Nevertheless, a U.S. national survey in 2009 suggested that a third of the U.S. population had heard nothing about nanotechnology (Hart Research 2009), and a 2009 meta-analysis of survey studies conducted between 2002 and 2007 (most of which were conducted in North America) estimated that 51% of respondents had heard nothing about nanotechnology (Satterfield et al. 2009). This absence of awareness is generally confirmed by other studies (e.g., Besley 2010). Although a 2011 survey in Australia suggested a possible trend toward increased awareness, of the 76% who affirmed (when prompted) that they had heard of the term *nanotechnology* (51% in 2005), the majority reported they didn't know what it means or didn't know how it works. A scant one-third (29%) of respondents volunteered awareness of any products that include or are made with nanotechnology (Australian Govt. 2011b).

In general, nano risk–benefit attitudes absent much knowledge are driven by heuristic, affective responses to nanotechnology (Kahan et al. 2007) and to science and technology generally (Satterfield et al. 2009); they are constructed in response to the specific context (Slovic 1995). Cultural attitudes, for example, predict familiarity with nanotechnology (Kahan et al. 2008a) and the extent to which people feel nanotechnology "has or has not been developed with their values, interests and beliefs in mind" (Priest 2006). When people receive new information on nanotechnology, risk attitudes can sometimes polarize (Kahan et al. 2008b), contingent on the similarity the recipient perceives between the spokesperson's values and his or her own. So trust is of central importance in nanotechnology risk communication (Satterfield et al. 2009), as in other risk communication contexts (Cvetkovich and Löfstedt 1999; Earle and Cvetkovich 1995; Löfstedt 2005; Siegrist, Cvetkovich, and Roth 2000; Poortinga and Pidgeon 2006).

In one risk perception study that addressed nano applications in cosmetics, the vast majority of (Norwegian) consumers in the study were unaware of specific nano applications (Throne-Holst and Strandbakken 2009). In their focus group of consumers, participants expressed fear about the use of nanotechnology in everyday products. One woman in the group asked: "How can it be possible that these things are being produced when their potential harmful effects are unknown?" These consumers were surprised to be told that it is arguably not currently possible to satisfactorily judge the risk of nanomaterials in products. A male participant commented: "We trust the authorities; if it is on sale in a pharmacy, we regard it as safe" (Throne-Holst and Strandbakken 2009).

Yet, although Australian governmental reviews (2006, 2009) characterized nano-sunscreens as safe, almost one-fifth of respondents in a 2011 Australian survey reported that they would go out of their way to avoid using a sunscreen that had nanoparticles in it, and about half said that they would need to know more about it before deciding (Australian Govt. 2012). This result echoes a dominant theme from a national consensus conference on nanotechnology in the United States (Hamlet et al. 2008) as summarized by Philbrick and Barandiaran (2009): "If there is one area of agreement among [National Citizens' Technology Forum] participants nationwide, it is the need for governmental agencies to provide better information to the public."

Although awareness of nanotechnology is generally low, people still apply some mental model to make inferences about physical and social phenomena. As discussed previously, analogies can play an important role in this. In the absence of specific knowledge about or experience with a risk, judgments and decisions may be based on affective responses or trust, or rely on analogies and simple mental models (e.g., Bostrom 2008; see also Gentner and Stevens 1983; Gentner et al. 2001). The particular analogies people use focus their attention and influence their subjective risk assessments and other judgments and preferences regarding emerging nanotechnologies and nanomaterials.

Specific decision contexts trigger different judgments and preferences of nanotechnology (Siegrist et al. 2008; see also Slovic 1995). Although perceived benefits and technological enthusiasm continue to dominate nanotechnology perceptions and attitudes (for reviews see Besley 2010; Bostrom and Löfstedt 2010; Siegrist 2010), this is not uniform. Advocacy groups are divided, and in some contexts nanotechnology evokes a negative reaction among consumers (Siegrist, Stampfli, and Kastenholz 2009), as also discussed previously.

9.6 Conclusions

The European Chemicals Agency issued its first Community Rolling Action plan to assess the risks of suspected substances under REACH in

2012; it included two nanomaterials—silicon dioxide nanoparticles and silver nanoparticles (ECHA 2012a,b). As the shortcomings identified in this chapter highlight, there is clearly a need for more and better nanotechnology risk communication.

Lack of coordination and communication by government provides at least part of the explanation for the current state of affairs. The Inspector General's office at the U.S. EPA concluded at the end of 2011 that although the EPA has tried to communicate about nanomaterials and solicit input from stakeholders, "the agency as a whole has not provided a transparent overall message about nanomaterials to the general public" (US EPA 2011). To do this effectively will require risk communication science.

At the heart of risk communication science is the systematic study and understanding of how people think (and feel) about risk and exchange information about it (e.g., Fischhoff et al. 1978, 2011; Morgan et al. 2002; Slovic 1987, 2000). Among the facts in the field are that risk information is often processed heuristically, for example, based on a feeling of how bad or good a choice option is (Finucane et al. 2000), rather than on an analytic risk–benefit calculus. People often want or even need numbers, but many find numbers hard to understand, especially probabilities (Peters 2008). Faced with risk numbers, some people will rely on heuristics that make them susceptible to biases, depending on how the numbers are presented (Peters and Levin 2008). Numbers also tend to numb responses to the risk (Fetherstonhaugh et al. 1997). Risk messages that do evoke fear but fail to convey how to reduce or mitigate a risk can fall flat (Maloney, Lapinski, and Witte 2011). It may seem obvious, but risk sciences also demonstrate that people care about whether they have a choice (e.g., Fischhoff et al. 1978; Slovic 2000), and whether the values of others involved in managing and communicating the risk align with their own (Kahan et al. 2008a,b; Cvetkovich and Löfstedt 1999; Löfstedt 2005). All of this is relevant to nanotechnology risk communication; some but not all of it has been studied in this context.

Despite the growing body of evidence regarding nanotechnology risk perceptions, empirical evaluation of nanotechnology risk communications is essential (Fischhoff, Brewer, and Downs 2011). Relatively inexpensive, pragmatic approaches to evaluating risk messages and programs are a start; government agencies and other organizations producing nanotechnology risk communications would do well to invest in them (Fischhoff, Brewer, and Downs 2011).

References

Australian Govt. 2011a. Notes of a discussion with Dr Ken Harvey, Transparency Review Panel. Department of Health and Aging, Therapeutic Goods Administration (TGA). Transparency Public Consultation, meeting with Friends of the Earth, Nanotechnology Project Personnel, Brunswick, March 3, p. 1. http://www.tga.gov.au/pdf/submissions/review-tga-transparency-1101-submission-foe.pdf

Australian Govt. 2011b. Australian community attitudes held about nanotechnology: Trends2005to2011.MarketAttitudeResearchServicesPty.Ltd.http://www.innovation.gov.au/Industry/Nanotechnology/PublicAwarenessandEngagement/Documents/NanotechnologyPublicAwareness2011.pdf

Australian Govt. 2012. Study of public attitudes towards sunscreens with nano particles. Department of Industry, Innovation, Science, Research and Tertiary Education. http://www.innovation.gov.au/Industry/Nanotechnology/Public AwarenessandEngagement/Documents/SunscreenStudy.pdf

Bellucci, S., and R. Burri. 2008. Public perception of nanotechnology. *J. Nanoparticle Res.* 10(3): 387–91.

Besley, J. 2010. Current research on public perceptions of nanotechnology. *Emerging Health Threats J.* 3:e8 (online); doi:10.3134/ehtj.10.164

Bostrom, A. 2008. Lead is like mercury: Risk comparisons, analogies and mental models. *J. Risk Res.* 11(1–2): 99–117.

Bostrom, A., and R. E. Löfstedt. 2010. Nanotechnology risk communication past and prologue. *Risk Analysis* 30(11): 1645–62.

Bowman, D. M., and G. A. Hodge. 2007. Nanotechnology and public interest dialogue: Some international observations. *Bull. Sci. Technol. Soc.* 27(2): 118–32.

Christensen, F. M., H. J. Johnston, V. Stone, R. J. Aitken, S. Hankin, S. Peters, and K. Aschberger. 2011. Nano-TiO$_2$: Feasibility and challenges for human health risk assessment based on open literature. *Nanotoxicology* 5(2): 110–24.

Consumers Union. 2008. Consumers Union urges FDA to conduct full scale safety review of nanoparticles in sunscreen: Consumer Reports test reveals nanoparticles present in 4 out of 5 sunscreens, even when companies claim they are not. Yonkers, NY: Consumers Union. http://www.consumersunion.org/pub/core_product_safety/006259.html

Cook, A. J., and J. R. Fairweather. 2005. Nanotechnology—Ethical and social issues: Results from New Zealand focus groups. Research report No. 281. Canterbury, New Zealand: Agribusiness and Economics Research Unit, Lincoln University. http://citeseerx.ist.psu.edu/viewdoc/summary?doi=10.1.1.135.482

Cvetkovich, G., and R. E. Löfstedt, eds. 1999. *Social trust and the management of risk.* London: Earthscan.

Devries, S. P., and S. Jehl. 2009. Nanotechnology: The next battleground for mass torts? *Toxics Law Reporter* 24(47): 11362–64. http://www.winston.com/siteFiles/Publications/PDFArticleBNA.pdf

Drexler, K. E. 1986. *Engines of Creation: The Coming Era of Nanotechnology.* New York: Anchor Books.

Earle, T. C., and G. T. Cvetkovich. 1995. *Social Trust: Towards a Cosmopolitan Society.* London: Praeger.

ECHA. 2012a. Substance evaluation starts under REACH: The first list of substances published. ECHA/PR/12/05. Press release. European Chemicals Agency Press, Helsinki. http://echa.europa.eu/web/guest/view-article/-/journal_content/c26e0b90-8d88-4580-9954-842a934486a1

———. 2012b. ANNEX to the opinion of the member state committee on the first draft Community Rolling Action Plan (CoRAP). European Chemicals Agency Press, Helsinki. http://echa.europa.eu/documents/10162/13578/final_annex_to_msc_opinion_on_corap_2012_public_en.pdf

ETC Group. 2005. A tiny primer on nano-scale technologies...and the Little BANG Theory. Action Group on Erosion, Technology and Concentration, Ottawa, Canada. www.etcgroup.org/upload/publication/55/01/tinyprimer_english.pdf

Fetherstonhaugh, D., P. Slovic, S. M. Johnson, and J. Friedrich. 1997. Insensitivity to the value of human life: A study of psychophysical numbing. *J. Risk Uncertainty* 14(3): 283–300.

Finucane, M. L., A. S. Alhakami, P. Slovic, and S. M. Johnson. 2000. The affect heuristic in judgments of risk and benefits. *J. Behav. Dec. Making* 13: 1–17.

Fischhoff, B., N. Brewer, and J. S. Downs, eds. 2011. *Communicating risks and benefits: An evidence-based user's guide.* Washington, DC: Food and Drug Administration.

Fischhoff, B., P. Slovic, S. Lichtenstein, S. Read, and B. Combs. 1978. How safe is safe enough? A psychometric study of attitudes towards technological risks and benefits. *Policy Sciences* 9(2): 127–52.

FOE. 2008. Out of the laboratory and onto our plates: Nanotechnology in food and agriculture. Report prepared for Friends of the Earth Australia, Europe, and United States, and supported by Friends of the Earth Germany. Written by G. Miller and Dr. R. Senjen, Friends of the Earth Australia Nanotechnology Project, 2nd ed.

Friedman, S. M., and B. P. Egolf. 2005. Nanotechnology: Risks and the media. *IEEE Tech. Soc. Mag.* 24(4): 5–11.

———. 2011. A longitudinal study of newspaper and wire service coverage of nanotechnology risks. *Risk Analysis* 31(11): 1701–17.

Gentner, D. 2002. Mental models, Psychology of. In *International Encyclopedia of the Social and Behavioral Sciences*, ed. N. J. Smelser and P. B. Bates, 9683–87. Amsterdam: Elsevier Science.

Gentner, D., K. J. Holyoak, and B. Kokinov, eds. 2001. *The Analogical Mind: Perspectives From Cognitive Science.* Cambridge, MA: MIT Press.

Hackenberg, S., G. Friehs, M. Kessler, K. Froelich, C. Ginzkey, C. Koehler, A. Scherzed, M. Burghartz, and N. Kleinsasser. 2011. Nanosized titanium dioxide particles do not induce DNA damage in human peripheral blood lymphocytes. *Environ. Molec. Mutagenesis* 52(4): 264–68.

Hamlet, P., M. B. Cobb, and D. H. Guston. 2008. National Citizens Technology Forum: Nanotechnologies and human enhancement. CNS-ASU report No. R08-0003. Center for Nanotechnology in Society, Arizona State University, Tempe, AZ.

Hart Research Associates. 2009. Nanotechnology, synthetic biology, and public opinion. Report for the Project on Emerging Nanotechnologies. The Woodrow Wilson International Center For Scholars, Washington DC, September 22.

Holdren, J. P., C. R. Sunstein, and I. A. Siddiqui. 2011. Memorandum for the heads of executive departments and agencies on policy principles for the U.S. decision-making concerning regulation and oversight of applications of nanotechnology and nanomaterials. Executive Office of the President: Office of Management and Budget, U.S. Trade Representative, Office of Science and Technology Policy. http://www.whitehouse.gov/sites/default/files/omb/inforeg/for-agencies/nanotechnology-regulation-and-oversight-principles.pdf

Kahan, D., P. Slovic, D. Braman, J. Gastil, and G. Cohen. 2007. Affect, values, and nanotechnology risk perceptions: An experimental investigation. Cultural Cognition Project Working Paper No. 22. New Haven, CT: Yale Law School. http://papers.ssrn.com/sol3/papers.cfm?abstract_id=968652

———. 2008a. The future of nanotechnology risk perceptions: An experimental investigation. New Haven, CT: Yale Law School. http://papers.ssrn.com/sol3/papers.cfm?abstract_id=1089230

Kahan, D., P. Slovic, D. Braman, J. Gastil, G. Cohen, and D. Kyser. 2008b. Biased assimilation, polarization, and cultural credibility: An experimental study of nanotechnology risk perceptions. New Haven, CT: Yale Law School. http://ssrn.com/abstract = 1090044

Kennedy, J. F. 1962. Special message to the Congress on protecting the consumer interest. March 15. The American Presidency Project, ed. J. T. Woolley and G. Peters. http://www.presidency.ucsb.edu/ws/?pid=9108

Kleinman, D., and M. Powell. 2005. Report of the Madison Area Citizen Consensus Conference on Nanotechnology. http://www.nsec.wisc.edu/NS-Nugget.php?ID = 3

Löfstedt, R. E. 2005. *Risk Management in Post Trust Societies*. Basingstoke, England: Palgrave/Macmillan.

Maloney, E. K., M. K. Lapinski, and K. Witte. 2011. Fear appeals and persuasion: A review and update of the extended parallel process model. *Social Personality Psychol.* 5(4): 206–19.

Maynard, A. 2007. Nanotechnology: A primer. Woodrow Wilson Center, Washington DC. http://www.nanotechproject.org/publications/archive/nanotechnology_primer/

———. 2009a. Nanotechnology: From nano-novice to nano-genius in 13 steps. *2020 Science*. http://2020science.org/2009/05/26/nanotechnology-primer/

———. Maynard, A. 2009b. Managing the small stuff: A visual nanotechnology primer. *2020 Science*. http://2020science.org/2009/02/02/managing-the-small-stuff-a-visual-nanotechnology-primer/

McCully, S. 2009. Defining natural: A new standard for body care products. PCC Natural Markets, Seattle, WA. http://www.pccnaturalmarkets.com/sc/0905/sc0905-natural.html

Morgan, M. G., B. Fischhoff, A. Bostrom, and C. J. Atman. 2002. *Risk communication: A mental models approach*. Cambridge, England: Cambridge University Press.

Nersessian, N. J., and S. Chandrasekharan. 2009. Hybrid analogies in conceptual innovation in science. *Cognitive Systems Res. J.* 10(special issue): 178–88.

Nisbet, M. C., and D. A. Scheufele. 2007. The future of public engagement. *Scientist* 21(10): 38–44.

Osborne, L. 2003. The gray-goo problem. *New York Times Magazine*, December 14. http://www.nytimes.com/2003/12/14/magazine/14GRAY.html

PCC. 2011. Summary of issues before the NOSB (National Organic Standards Board). Puget Consumer Cooperative, Seattle, WA. http://www.pccnatural-markets.com/events/nosb2011_topics.html

Peters, E. 2008. Numeracy and the perception and communication of risk. *Ann. N.Y. Acad. Sci.* 1128: 1–7.

Peters, E., and I. P. Levin. 2008. Dissecting the risky-choice framing effect: Numeracy as an individual-difference factor in weighting risky and riskless options. *Judgment Decision Making* 3(6): 435–48.

Philbrick, M., and J. Barandiaran. 2009. The National Citizens' Technology Forum: Lessons for the future. *Science Public Policy* 36(5): 335–47.

Pidgeon, N., B. Herr Harthorn, K. Bryant, and T. Rogers-Hayden. 2009. Deliberating the risks of nanotechnologies for energy and health applications in the United States and United Kingdom. *Nature Nanotechnol.* 4:95–98

Pidgeon, N., and T. Rogers-Hayden. 2007. Opening up nanotechnology dialogue with the publics: Risk communication or "upstream engagement"? *Health Risk Society* 9(2): 191–210.

Poland, C. A., R. Duffin, I. Kinloch, A. Maynard, W. A. H. Wallace, A. Seaton, V. Stone, S. Brown, W. MacNee, and K. Donaldson. 2008. Carbon nanotubes introduced into the abdominal cavity display asbestos-like pathogenic behaviour in a pilot study. *Nature Nanotechnol.* 3:423–28.

Poortinga, W., and N. F. Pidgeon. 2006. Prior attitudes, salient value similarity, and dimensionality: Toward an integrative model of trust in risk regulation. *J. Applied Social Psychol.* 36(7): 1674–1700.]

Priest, S. 2006. The North American opinion climate for nanotechnology and its products: Opportunities and challenges. *J. Nanoparticle Res.* 8:563–68.

Ruivenkamp, M., and A. Rip. 2010. Visualizing the invisible nanoscale study: Visualization practices in nanotechnology community of practice. *Science Studies* 23(1): 3–36.

Safe Work Australia. 2010. An evaluation of MSDS and labels associated with the use of engineered nanomaterials. http://www.safeworkaustralia.gov.au/ABOUTSAFEWORKAUSTRALIA/WHATWEDO/PUBLICATIONS/Pages/RP201006EvaluationOfMSDSAndLabels.aspx

Salamon, A. W., P. Courtney, and I. Shuttler. 2010. Frequently asked questions: Nanotechnology and engineered nanomaterials: A primer. PerkinElmer, Waltham, MA. http://shop.perkinelmer.com/Content/Manuals/GDE_NanotechnologyPrimer.pdf

Sargent Jr., J. F. 2011. Nanotechnology: A policy primer. U.S. Congressional Research Service 7-5700, RL34511. http://fpc.state.gov/documents/organization/137018.pdf

Satterfield, T., M. Kandlikar, C. E. H. Beaudrie, J. Conti, and B. H. Harthorn. 2009. Anticipating the perceived risk of nanotechnologies. *Nature Nanotechnology* 4(11): 752–58.

Scheufele, D. A. 2006. Manufacturers removing nano labels from product descriptions. *Nanopublic: public opinion, media, and emerging (communication) technologies.* http://www.nanopublic.com/2006/11/manufacturers-removing-nano-labels.html

Sharifi, S., S. Behzadi, S. Laurent, M. Laird Forrest, P. Stroeve, and M. Mahmoudi. 2012. Toxicity of nanomaterials. *Chem. Soc. Rev.* 41: 2323–43.

Shatkin, J. A., L. C. Abbott, A. E. Bradley, R. A. Canady, T. Guidotti, K. M. Kulinowski, R. E. Löfstedt, et al. 2010. Nano risk analysis: Advancing the science for nanomaterials risk management. *Risk Analysis* 30(11): 1680–87.

Shatkin, J. A., and W. North. 2010. Perspectives on risks of nanomaterials and nanotechnologies: Advancing the science. *Risk Analysis* 30(11): 1627–33.

Siegrist, M. 2010. Predicting the future: Review of public perception studies of nanotechnology. *Human and Ecological Risk Assessment* 16(4): 837–46.

Siegrist, M., and K. Carmen. 2011. Labeling of nanotechnology consumer products can influence risk and benefit perceptions. *Risk Analysis* 31(11): 1762–69.

Siegrist, M., G. T. Cvetkovich, and C. Roth. 2000. Salient value similarity, social trust, and risk/benefit perception. *Risk Analysis* 20: 353–62.

Siegrist, M., N. Stampfli, and H. Kastenholz. 2009. Acceptance of nanotechnology foods: A conjoint study examining consumers' willingness to buy. *Brit. Food J.* 111(7): 660–68.

Siegrist, M., N. Stampfli, H. Kastenholz, and C. Keller. 2008. Perceived risks and perceived benefits of different nanotechnology foods and nanotechnology food packaging. *Appetite* 51(2): 283–90.

Singer, N. 2008. Bill seeks action on stricter sunscreen rules. *New York Times*, August 2.

Sladek. J. 1968. *The Reproductive System* (aka *Mechasm*). London: Gollancz.

Slovic, P. 1987. Perception of risk. *Science* 236(4799): 280–85.

———. 1995. The construction of preference. *Am. Psychologist* 50:364–71.

———. 2000. *The perception of risk.* London: Earthscan.

TA-Swiss. 2006. Public reactions to nanotechnology in Switzerland. Report on Publifocus Discussion Forum: Nanotechnology, Health and the Environment. TA-P 8/2006. Centre for Technology Assessment at the Swiss Science and Technology Council, Bern.

Throne-Holst, H., and P. Strandbakken. 2009. Nobody told me I was a nano-consumer: How nanotechnologies might challenge the notion of consumer rights. *J. Consumer Policy* 32:393–402.

Trouiller, B., R. Reliene, A. Westbrook, P. Solaimani, and R. H. Schiestl. 2009. Titanium dioxide nanoparticles induce DNA damage and genetic instability in vivo in mice. *Cancer Res.* 69(22): 8784–89.

U.S. EPA. 2011. EPA needs to manage nanomaterial risks more effectively. Report No. 12-P-0162. Office of Inspector General, Washington, DC.

10

Ongoing International Efforts to Address Risk Issues for Nanotechnology

Jo Anne Shatkin

CONTENTS

Publicly funded integrated research projects on a national and European level have concluded that a general "nano-risk" does not exist. Nano is nothing new.

Dr. Andreas Kreimeyer
BASF

The explosive growth of nanotechnology and its potential penetration into so many sectors of the economy have prompted broad international efforts to address the issues of occupational, public health, and environmental risks. Numerous organizations—governmental, nongovernmental, professional, not for profit—have developed nanotechnology activities relating to its environmental, legal, societal, and ethical impacts. Much of this activity involves intergovernmental collaboration, academic liaisons, and other associations, and also includes entities developing voluntary standards. The large number of efforts internationally means only select efforts will be mentioned here. Many of these organizations and efforts are mentioned in other parts of the book, but are consolidated here to provide a fairly comprehensive assessment.

Nanotechnology is not as widely discussed as it was in the mid-2000s, when many nongovernmental organizations were publishing reports, petitioning agencies to regulate nanomaterials, and staging protests. (Two such events included a naked group of activists disrupting a nanotechnology meeting and an Eddie Bauer store window in Chicago protesting the use of nanotechnology in pants [Lovy 2005], and another by the "Yes Men" who infiltrated an international technical nanotechnology conference in 2005 in San Francisco, pretending to be from the ethics division of Dow Chemical and espousing a "post-cautionary approach" to managing nanotechnology risks.)

In fact, despite the growth of governmental, industrial, and academic efforts to be responsive to public concerns and address the risks of engineered nanomaterials (ENM), many are suggesting that it is inappropriate to evaluate nanomaterials as a class, because particle size is insufficient to characterize the potential for risks and is a poor determinant of effects. The problem with not calling out size was already alluded to in Chapter 5, i.e., the difficulty in even identifying the amounts of ENM currently produced and which products contain them. Eventually, as we learn more about what makes some smaller particles hazardous, there will be ways to segment these out. Simply, the knowledge base remains too sparse to stop drawing bounds around the nanoscale for regulatory consideration.

The efforts reported on in this chapter generally include a risk component. These are the organizations, or groups of organizations, that are contributing to the international dialogue on how to identify, assess, and manage the environmental health and safety aspects of nanomaterials and nanotechnology, and the list is not inclusive. The focus here is more on environmental aspects and less on occupational exposures, which are addressed in Chapter 8.

Since 2008, when the first edition of this book was published, many more governmental efforts to address the risks of nanomaterials and nanotechnologies have developed. Yet the regulatory environment continues to be dynamic, and that brings challenges for producers, regulators, and the public to keep up to date on activities.

10.1 International Federal Governmental Efforts

Australia's National Industrial Chemicals Notification and Assessment Scheme (NICNAS) published new requirements for the notification of nano-forms of chemicals not listed on the Australian Inventory of Chemical Substances (NICNAS 2011). Adjustments to NICNAS's New Chemicals Program processes mean that all industrial nanomaterials will undergo a premarket assessment by NICNAS. Similar to policies in the United States and Canada, nano-forms of chemicals already listed on the Australian Inventory of Chemical Substances are considered to be existing chemicals and, at present, can be legally introduced and used in Australia without notification to NICNAS. Additional information may be required on a case-by-case basis. There are some key definitions used by NICNAS (2011) for defining industrial nanomaterials.

NICNAS Working Definition of Industrial Nanomaterial

...industrial materials intentionally produced, manufactured or engineered to have unique properties or specific composition at the nanoscale, that is a size range typically between 1 nm and 100 nm, and is either a nano-object (i.e., that is confined in one, two, or three dimensions at the nanoscale) or is nanostructured (i.e. having an internal or surface structure at the nanoscale).

[Notes to the working definition:

- intentionally produced, manufactured or engineered materials are distinct from accidentally produced materials
- "unique properties" refers to chemical and/or physical properties that are different because of their nanoscale features as compared to the same material without nanoscale features, and result in unique phenomena (e.g., increased strength, chemical reactivity, or conductivity) that enable novel applications
- aggregates and agglomerates are considered to be nanostructured substances
- where a material includes 10% or more number of particles that meet the above definition (size, unique properties, intentionally produced) NICNAS will consider this to be a nanomaterial]" (NICNAS 2011)

Food Standards Australia New Zealand (FSANZ) developed a fact sheet on nanotechnology, explaining that:

While the term may be new, food and water are naturally made up of nm-scale particles. For example, proteins are in the nanoscale size range and milk is an emulsion of nanoscale fat droplets.... Any new

> foods manufactured using nanotechnologies that may present safety
> concerns will have to undergo a comprehensive scientific safety
> assessment before they can be legally supplied in Australia and New
> Zealand.… FSANZ has not received any applications to approve new
> or novel nanoscale particles for food use. (FSANZ 2011)

China is investing in basic scientific research on the biological interactions of nanomaterials at the cellular and organ level in order to establish safety standards. Research includes efforts to model the behavior of nanoparticles and building a database of properties and effects of several nanomaterials in order to establish safe approaches for managing "artificial nano-materials," which is considered part of maintaining a competitive edge (Chinese Academy of Sciences 2007).

In 2011, Health Canada published a working definition of manufactured nanomaterials for regulatory purposes to allow information gathering activities (Health Canada 2011).

> Health Canada considers any manufactured substance or product
> and any component material, ingredient, device, or structure to be
> nanomaterial if:
>
> a. It is at or within the nanoscale in at least one external dimension, or has internal or surface structure at the nanoscale, or;
> b. It is smaller or larger than the nanoscale in all dimensions and exhibits one or more nanoscale properties/phenomena.
>
> For the purposes of this definition:
> i. The term "nanoscale" means 1 to 100 nanometres, inclusive;
> ii. The term "nanoscale properties/phenomena" means properties which are attributable to size and their effects; these properties are distinguishable from the chemical or physical properties of individual atoms, individual molecules and bulk material; and,
> iii. The term "manufactured" includes engineering processes and the control of matter.

Health Canada and Environment Canada developed guidance for determining whether nanoscale materials are new or existing (EC 2007). Both agencies continue to actively review nanomaterial submissions for a breadth of consumer, medical, and food-related applications.

The Council of Canadian Academies convened an expert panel on health and environmental aspects of nanotechnology to address Health Canada's questions about the need to update their risk assessment approaches for nanotechnology (CCA 2009). The National Research Council of Canada continues its active research program. Alberta Innovates and the National

Institute for Nanotechnology also continue their leadership on sustainable applications of nanotechnology.

In Europe, the European Food Safety Commission released guidance in 2011 for assessing food and feed with engineered nanomaterials (ENM). If it is adequately demonstrated that the ENM do not reach the gastrointestinal tract, either because they dissolve or are not in contact with the food, then a traditional assessment is performed. If there is exposure to ENM, detailed physical chemical characterization protocols are used to measure properties, along with absorption, distribution, metabolism, and excretion studies. Depending on the scenario, subchronic toxicity studies may also be required (EFSA 2011). There have been numerous consultations in Europe on the regulation of nanomaterials and nanotechnologies under the Registration, Evaluation, and Authorization and Restriction of Chemicals (REACH), with a general view that for most applications, REACH is an adequate regulatory framework for ENM. As with most other regulatory agencies, the case-by-case approach highlights when the need for additional testing is required.

One significant development in Europe is the adoption by the European Commission of a controversial definition of nanomaterials (EC 2011d). The definition states:

> On 18 October 2011 the European Commission adopted the Recommendation on the definition of nanomaterial.
>
> According to this Recommendation:
>
> "Nanomaterial" means a natural, incidental or manufactured material containing particles, in an unbound state or as an aggregate or as an agglomerate and where, for 50% or more of the particles in the number size distribution, one or more external dimensions is in the size range 1–100 nm.
>
> In specific cases and where warranted by concerns for the environment, health, safety or competitiveness the number size distribution threshold of 50% may be replaced by a threshold between 1% and 50%.
>
> By derogation from the above, fullerenes, graphene flakes and single wall carbon nanotubes with one or more external dimensions below 1 nm should be considered as nanomaterials. (EC 2011d)

Many in industry and elsewhere have expressed concern about the breadth of this definition, because it suggests that nanomaterials will be identified by size alone, and not properties specific to the nanoscale, and will broadly encompass—and require additional regulatory consideration of—materials that have been in commerce for decades, including minerals, food additives, and other substances that are not novel.

In **Japan**, the Japanese government recently published risk assessments based on years of research on environmental health and safety aspects of nanotechnology (Nakanishi 2011). On the basis of these risk assessments, three Occupational Exposure Limits (OELs) are developed, described in Chapter 5. The recommendation is for a ten-year exposure limit, assuming that the limits will be reestablished in that time frame, given the emerging nature of our current understanding of exposure, toxicity, and risk. The government is convening a committee to review the risk assessments and develop a risk management strategy in 2012 (METI 2011).

In Korea, the Ministry of Knowledge and Economy (MKE) published a guideline on "Guidance on Safe Management of Nanotechnology Based Product." MKE is also working on phased development of a safe certification system for nano products that includes life cycle management, product registration, and certification (ANF 2011; OECD 2011).

10.1.1 Highlights of Governmental Activities in the United States

In the **United States**, efforts among several agencies in the federal government are ongoing. Federal efforts are coordinated through the National Nanotechnology Initiative (NNI). NNI is administered by the National Nanotechnology Coordination Office, in the White House Office of Science and Technology Policy, and oversees some $1.4 billion in funding for research and development of nanoscale technology. Roughly 10% of the proposed 2012 budget is expended on projects related to environmental health and safety (EHS) and to ethical, legal, and societal implications—although some within the NNI have suggested that this figure is an underestimate.

The U.S. government is engaged in several international efforts, with a specific focus on environmental health and safety (EHS). As detailed in the 2011 NNI EHS Strategy, the United States is participating in international standards work, round-robin testing, joint research funding, capacity building, and efforts through the Organization for Economic Cooperation and Development (OECD) to harmonize international efforts (NNI 2011).

In June 2011, the Office of Management and Budget, with the White House Office of Science and Technology Policy, released a memorandum about principles for regulatory decision making. The purpose is "promoting a balanced, science-based approach to regulating nanomaterials and other applications of nanotechnology in a manner that protects human health, safety, and the environment without prejudging new technologies or creating unnecessary barriers to trade or hampering innovation" (OMB 2011). The document describes principles for government agencies making judgments about nanomaterials, including,

> Nanomaterials should not be deemed or identified as intrinsically benign or harmful in the absence of supporting scientific evidence, and regulatory action should be based on such scientific evidence.

> In general, however, and to the extent consistent with law, regulation should be based on risk, not merely hazard.

Importantly, the principles state:

> For oversight and regulation, however, the critical issue is whether and how such new or altered properties and phenomena emerging at the nanoscale create or alter the risks and benefits of a specific application. A focus on novel properties and phenomena observed in nanomaterials may ultimately be more useful than a categorical definition based on size alone. Properties and phenomena emerging at the nanoscale enable applications that may alter the safety, effectiveness, performance, or quality of products — giving rise to both risks and benefits.

In 2009, the U.S.EPA concluded that their voluntary program for reporting on nanomaterials had largely failed, with only about twelve companies providing information. This led to U.S.EPA beginning to issue a series of policies, including guidance on new versus existing substances for the purpose of reporting under the Toxic Substances Control Act (TSCA) (U.S.EPA 2009). U.S.EPA has reviewed over 100 nanoscale submissions under the New Chemicals Program of TSCA, and uses a variety of tools to obtain information and approve submissions on a case-by-case basis. For example, U.S.EPA has issued Significant New Use Rules (SNURs) for several single-walled, double-walled, and other multiwalled carbon nanotubes and fullerenes (U.S.EPA 2011a, 2011b).

There is not much to say about one recent example, a multiwalled carbon nanotube that was the subject of a SNUR, because the entire file is considered confidential business information. We only know that it is for a nanotube because U.S.EPA states it is, but there is no information available about the manufacturer, how much is made, and all of the details of the testing (U.S.EPA 2011a). Table 10.1 shows a similar submission by another manufacturer (U.S.EPA 2011b). U.S.EPA requested comments on the draft SNUR, but what can one say about the data in Table 10.1, other than, "Hope you got it right, U.S.EPA." U.S.EPA's Office of Pesticide Programs (OPP) has also been working on assessing risks of nanoscale materials, primarily nanosilver. In 2011, U.S.EPA conditionally approved the first nanoscale silver antimicrobial submitted as such, for use in fabric treatment (according to Nowack et al. [2011]. U.S.EPA approved nanoscale colloidal silver as an antimicrobial in the 1950s).

In their risk assessment, U.S.EPA reviewed the literature and concluded that despite data gaps, the potential health and ecological risks were low. A tiered approach to filling the data gaps is part of the conditional registration (U.S.EPA 2011c). In response, the Natural Resources Defense

TABLE 10.1

Data for Public Comment on EPA Significant New Use Rule

Contains Confidential Business Information

Material Safety Data Sheet

Note: Table 10.1 is an example of a data sheet redacted by the U.S.EPA.

Council filed a lawsuit against U.S.EPA, charging that U.S.EPA does not have the authority to register the pesticide without the needed data and, by doing so, may create a condition of "unreasonable adverse effects on the environment" (NRDC 2010, 2012), despite OPP's conclusion that the conditional registration of the product does not pose unreasonable ecological or human health risk. NRDC is in part opposed because U.S.EPA uses the conditional registration status for several substances, raising concerns about new substances shortcutting the registration process (NRDC 2010).

A report by U.S.EPA's Office of Inspector General (U.S.EPA 2011d) found that "U.S.EPA does not currently have sufficient information or processes to effectively manage the human health and environmental risks of nanomaterials." The report notes that lack of formal internal coordination and, specifically, that the New Chemicals Program under TSCA relies on models, while the models have not been validated for nanomaterials.

The Food and Drug Administration prepared draft guidance for industry regarding applications with nanoscale materials under their governance:

> When considering whether an FDA-regulated product contains nanomaterials or otherwise involves the application of nanotechnology, FDA will ask:
>
> > Whether an engineered material or end product has at least one dimension in the nanoscale range (approximately 1 nm to 100 nm); or
> > Whether an engineered material or end product exhibits properties or phenomena, including physical or chemical properties or biological effects, that are attributable to its dimension(s), even if these dimensions fall outside the nanoscale range, up to one micrometer. (FDA 2011)

The guidance allows industry to prepare for additional data gathering on properties and behavior.

One activity in the United States is led by an interagency committee on Nanotechnology Environmental and Health Implications (NEHI). NEHI is part of the NNI and includes participants from the twenty-one agencies within the U.S. federal government that are responsible for management of nanotechnology EHS. These are: National Nanotechnology Coordination Office, Office of Science and Technology Policy, Office of Management and Budget, Consumer Product Safety Commission, Cooperative State Research Education and Extension Service, Department of Transportation, Food and Drug Administration, International Trade Commission, National Institute of Standards and Technology, Occupational Safety and Health Administration, National Science Foundation, Department of Defense, the Department of Energy, National Aeronautic Safety Administration,

National Institutes of Health, National Institute for Occupational Safety and Health, Department of Commerce, Department of Agriculture, U.S.EPA, Department of Justice, and the U.S. Geologic Survey.

Many other nations have national programs to develop nanotechnology and to ensure safe management of nanomaterials. One resource for these efforts is the Organization for Economic Cooperation and Development (OECD) Working Party on Nanomaterials (WPMN) that regularly publishes the recent governmental efforts of its members from their *Tour de Table* reports (OECD 2011).

10.2 Standard Setting

As described previously, there are national and international efforts to define nanomaterials for regulatory purposes, including efforts to harmonize measurements and testing requirements. Still, the regulatory situation is dynamic, as agencies sort out how much additional effort is required to conduct risk assessment for engineered nanomaterials and nano-enabled products. With a limited ability to identify products that contain ENM, and a recognition by some agencies, such as U.S.EPA, that there are likely products on the market that contain engineered nanomaterials, but have not been registered as such, agencies are evaluating alternatives for capturing nano-enabled materials and ensuring that they are adequately tested. For example, U.S.EPA's Office of Pesticide Programs is considering using clause 6(a)2 under the Federal Insecticide, Fungicide, and Rodenticide Act to require all submitters to inform U.S.EPA whether products contain nanomaterials. The Canadian government was planning on a data call for all products using nanomaterials to report their use in Canada in 2009, but has not issued the call as this book goes to press.

In the European Union, reporting rules are requiring actions regarding nanomaterials in certain products. For example, in 2013, the Cosmetics Directive will be replaced by Regulation EC 1223/2009, which requires labeling and reporting of data about ingredients including nanomaterials in cosmetics (EC 2011a). Directive 2011/65/EU (EC 2011b) directs:

> As soon as scientific evidence is available, and taking into account the precautionary principle, the restriction of other hazardous substances, including any substances of very small size or with a very small internal or surface structure (nanomaterials) which may be hazardous due to properties relating to their size or structure, and their substitution by more environmentally friendly alternatives which ensure at

least the same level of protection of consumers should be examined. To this end, the review and amendment of the list of restricted substances in Annex II should be coherent, maximise synergies with, and reflect the complementary nature of the work carried out under other Union legislation, and in particular under Regulation (EC) No. 1907/2006 while ensuring the mutually independent operation of this Directive and that Regulation.

This was a relief to many who feared, based on earlier committee recommendations, that nanosilver and carbon nanotubes would be specifically listed in Annex II. However, a new regulation—No. 1169/2011, on the provision of food information to consumers—requires labeling of foods containing engineered nanomaterials specifically listed as (nano) in the ingredient list (EC 2011b). This amended regulation is likely to cause confusion because the definition in it differs from the recommended definition for nanomaterials discussed on page 235. If the food labeling regulation adopted the recommended definition of a nanomaterial, most food ingredients would wrongly be listed as "nano." Instead, the "engineered" nanomaterial definition differs from, but is consistent with, other internationally accepted definitions for ENM that use both size and "unique property" criteria.

A snapshot of legislative updates presents the picture from 2009. In a survey of OECD member countries, seventeen regulatory efforts were identified that address notification, consumer and worker protection, environment, and importation and disposal of nanomaterials (OECD).

10.2.1 Voluntary Standards

Despite these regulatory developments, the situation is still evolving as regulators make decisions on a case-by-case basis. The nature of nanomaterial and nanotechnology risk is dynamic and evolving, requiring considerable diligence to keep up-to-date. In response, a number of organizations are developing voluntary standards for ENM, nano-objects, and products of nanotechnology. The purpose of voluntary standards is to establish rules of practice, for example, what to call certain materials, how to measure them, and how to protect workers amid uncertainty about risks. Some private companies are setting internal standards; these may be for worker protection, internal policies for product development, or data requirements for their supply chain. Others are multi-sector organizations that generally require membership to participate in standard setting and gain access to the standards, but tend to be open to participants from various sectors.

The premier organization developing voluntary standards for nanomaterials is the International Organization for Standards (ISO). ISO is addressing terminology, characterization of materials, and environmental safety and health. There is a nanotechnology committee within ISO, Technical Committee 229 (TC229), that is developing several voluntary standards for terminology, characterization, and risk assessment, among others. ISO has national committees that take on the development of specific standards that are brought to ISO for consideration. The American National Standards Institute (ANSI), the U.S. Technical Advisory Group to ISO, is leading the coordination of the environmental safety and health standards, including a standard for physical chemical characterization of nanomaterials for toxicity testing (ISO 2012). The European Committee on Standardization (CEN) led the development of the ISO terminology standard and several toxicity tests, and it continues efforts on others (CEN 2011).

As of 2011, over twenty standards had been published by TC229, including several terminology standards, measurement standards, and guidance for risk assessment (a "technical report") that incorporates adaptive management and life cycle thinking, similar to the Nano Risk Framework used by DuPont and developed collaboratively with EDF, with many attributes similar to Nano LCRA.

The technical report, "Nanotechnologies: Nanotechnology Risk Evaluation" (ISO/TR 13121:2011) is more of a guidance document than a standard, reflecting the uncertainties in the current state of knowledge about methods for assessing risks. That is, in order to develop standards, even "voluntary" ones, there must be considerable understanding and agreement about the nature of the hazard and the measurements for risk. The diversity of nanomaterials and nano-objects and applications precludes a governing "standard" approach at this point in time, mainly because of the physical impacts of particles, which is different from, and potentially additive to, assessing risks from conventional chemical substances.

There are other international standards organizations focused on nanotechnology standards, including ASTM International, Institute of Electrical and Electronics Engineers (IEEE), and the International Electrotechnical Commission (Technical Committee 113).

One effort I am involved in is the Nanorelease Project. A multisector international working group is coordinating efforts to develop standard methods for measuring the release of nanomaterials from products. The work is important because of all the categories of those potentially exposed: workers, consumers, and the environment. Say, for example, carbon nanotubes are used for flame retardant furniture. Over time, the furniture will wear. As we have discussed in earlier chapters, the need to consider these exposure over the life cycle is critical. An urgent need for producers, advocates, and regulators is to have replicable measurement methods (Nanorelease 2012).

10.2.2 Organization for Economic Cooperation and Development

The Organization for Economic Cooperation and Development (OECD) Working Party on Manufactured Nanomaterials (WPMN) is part of the OECD chemicals committee and promotes "international cooperation on human health, and environmental safety of manufactured nanomaterials, and involves approaches to safety testing and risk assessment of manufacturing nanomaterials" (OECD 2011). Governmental activities are coordinated by convening groups to discuss and agree upon a research agenda, coordinating efforts to ensure that research funding is leveraging the efforts across agencies. The main areas of focus are: identification and characterization, including terminology and standards; testing methods; risk assessment; and information sharing and dissemination (OECD 2007).

The OECD develops and publishes a variety of methods for chemical substances and, now, for nanomaterials. For example in the EU, OECD test methods are used to report chemical properties and toxicity of various end points under Registration, Evaluation, and Authorization of Chemicals (REACH). The OECD WPMN has also developed standardized methods for measuring nanomaterial properties (OECD 2010a) and is currently applying the existing test methods to nanomaterials in a round-robin testing program called the OECD Sponsorship Programme for the Testing of Manufactured Nanomaterials. Participants in OECD, including national governments and industry organizations, have sponsored the testing of specific substances to build a database of information about the toxicity and environmental behavior of these nanomaterials (OECD 2007, 2010b). Table 10.2 is a matrix of materials and tests that are currently sponsored. The standardization of test methods allows comparison of different nanomaterials and is anticipated to make large strides toward standardization of approaches for toxicity testing for nanomaterials. The current view is that the methods are applicable:

> The applicability of the OECD Test Guidelines for testing manufactured nanomaterials has been reviewed by the OECD Working Party on Manufactured Nanomaterials. This review found that in general the OECD Test Guidelines are applicable for investigating the health effects of nanomaterials, although it was noted that in some cases there will be a need for a further modification to the OECD guideline. This particularly applies to studies using the inhalation route and to toxicokinetic studies. [NICNAS 2011]

However, it is anticipated that some methods may be problematic and need to be modified. For example, as discussed in earlier chapters, there is poor correlation between *in vitro* and *in vivo* toxicity tests. One reason is the binding activity of certain nanomaterials to test agents or components

TABLE 10.2

OECD WPMN Nanomaterial Testing Program Substances and Endpoints

Nanomaterials	Nanomaterial Information/ Identification	Physical-Chemical Properties and Material Characterization	Environmental Fate	Environmental Toxicology	Mammalian Toxicology
Fullerenes (C60)	Nanomaterial name (from list)	Agglomeration/ aggregation	Dispersion stability in water	Effects on pelagic species	Pharmacokinetics (ADME)
Single-walled carbon nanotubes (SWCNTs)	CAS Number	Water solubility	Biotic degradability	Effects on sediment species	Acute toxicity
Multi-walled carbon nanotubes (MWCNTs)	Structural formula/ molecular structure	Crystalline phase	Ready biodegradability	Effects on soil species	Repeated dose toxicity
Silver nanoparticles	Composition	Dustiness	Surface water degradation	Effects on terrestrial species	If available
Iron nanoparticles	additives	Crystallite size	Soil simulation testing	Effects on microorganisms	Chronic toxicity
Carbon black	Basic morphology	Representative TEM picture(s)	Sediment simulation testing	Other relevant information	Reproductive toxicity
Titanium dioxide	Surface chemistry	Particle size distribution	Sewage treatment simulation testing		Developmental toxicity
Aluminium oxide	Major commercial uses	Specific surface area	Identification of degradation product(s)		Genetic toxicity

Cerium oxide	Known catalytic activity	Zeta potential (surface charge)	Degradation product testing	Experience with human exposure
Zinc oxide	Method of production	Surface chemistry (where appropriate)	Abiotic degradability and fate	Other relevant test data
Silicon dioxide		Material Safety	Hydrolysis, for surface modified nanomaterials	
Polystyrene		Flammability	Adsorption- desorption	
Dendrimers		Explosivity	Adsorption to soil or sediment	
Nanoclays		Incompatibility	Bioaccumulation potential	

of test media, which interferes with the measurement of specific end points (Stone, Johnston, and Schins 2009).

There are industrial members of OECD WPMN—including the U.S.-based American Chemistry Council Nanotechnology Panel; the Japan Chemical Industry Association; and CEFIC, the European Chemical Industry Council under the leadership of the Business and Industry Advisory Council (BIAC)—that provide guidance to members and others. For example, CEFIC provided a guide on how to manage nanomaterials under the Responsible Care program (OECD 2011).

The Nanotechnologies Industry Association (NIA) also participates in the OECD WPMN and conducts research and advocacy on behalf of its members. The NIA is the lead sponsor on testing of zinc oxide in the OECD sponsorship program (OECD 2011), and is participating in methods development and testing of cerium oxide as well. NIA participates in the CEN TC352 (CEN 2011) and the British Standards Institute nanotechnology standards development. One of their key roles has been in the development and implementation of the Responsible Nanocode, a best practices set of principles for those who sign on to them in the absence of clear regulatory guidance. The Nanocode has been adopted by a few proactive organizations, like BASF, Johnson and Johnson, and Unilever, but has not seen wide adoption by corporations.

10.3 Professional Organizations

There are several professional societies with specialty groups in ENM, nanotechnology and EHS, including Society for Risk Analysis (SRA), Society for Toxicology, Society for Environmental Toxicology and Chemistry (SETAC) and the American Chemical Society (ACS), IEEE, and others. In December 2006, I led the organization of the Emerging Nanoscale Materials Specialty (ENMS) Group of the SRA. The group currently has over 130 members representing government, academia, industry, and nonprofit organizations in twenty-two countries. Emerging nanoscale materials are agents recently identified or created that, as we have found, confer unique properties due to small size. The overarching goals of the group are:

> To facilitate the exchange of ideas and knowledge among practitioners, researchers, scholars, teachers, and others interested in risk analysis and emerging nanoscale materials

To encourage collaborative research on risk analysis and emerging nanoscale materials

To provide leadership and play an active role in advancing issues related to risk analysis and emerging nanoscale materials

ENMS is actively developing collaborative efforts with other groups within and outside of the SRA (SRANANO.org) and published a special collection of papers in SRA's journal *Risk Analysis* in 2010 (Shatkin and North 2010), stemming from an international workshop on nanotechnology and risk held in 2008. SRA was formed in 1980 and is an interdisciplinary international organization, an open forum for anyone interested in risk analysis. With hundreds of members internationally in local sections and chapters, as well as in specialty groups, SRA provides a home to analysts, communicators, decision makers, and others. SRA publishes the journal *Risk Analysis* and hosts annual meetings, conferences, and workshops on topics of risk. SRA ENMS recently provided expert consultation to the OECD WPMN subgroup 6 on risk assessment, convening experts to weigh in on the Critical Issues in Risk Assessment for Nanomaterials (SRA 2012).

The Society for Toxicology (SOT) has a specialty section on nanotoxicology (SOT 2007) and collaborates with others on symposia and educational efforts. For example, SRA, SOT, and the American Chemical Society cosponsored a congressional briefing on nanotechnology in 2009. Other groups include the Society for Environmental Toxicology and Chemistry (SETAC), ACS, International Association of Nanotechnology (IANANO), and others.

SETAC is also active in addressing nanotechnology and environmental issues each year in their annual meeting, where numerous abstracts and papers are presented on related topics. SETAC has also organized international efforts on life cycle analysis and its application, which includes nanotechnology.

The ACS meets semiannually and provides a forum for chemists to discuss all aspects of nanotechnology, including environmental aspects. *Chemical and Engineering News*, a weekly publication, once provided an annual report on nanotechnology but now reports weekly with updates on developments, in addition to covering news and other events and developments.

The Materials Research Society hosts semi-annual meetings with a large focus on nanotechnology and publishes research reports. The International Association of Nanotechnology (IANANO) is a multidisciplinary organization that promotes research and business development for the nanotechnology industry, and hosts annual conferences on nanotechnology. As mentioned previously, IEEE has an active nanotechnology group and publishes standards and provides other resources for their members.

10.4 Nongovernmental Organizations Addressing Environmental and Risk Issues

The Foresight Nanotech Institute is among the oldest of the nanotechnology organizations. Its mission is to enhance the beneficial implementation of nanotechnology and seek to guide nanotechnology research public policy and education around six major challenges. The challenges include: providing renewable clean energy; supplying clean water globally; improving health and longevity; healing and preserving the environment; making information technology available to all; and enabling space development (Foresight 2012).

The Woodrow Wilson International Center for Scholars (WWCS) Project on Emerging Nanotechnologies was a prime contributor to public knowledge about nanotechnology, developing databases, holding workshops, commissioning reports, and conducting surveys, all while funded by the Pew Charitable Trust. PEN is largely inactive now, although it continues to update its widely cited consumer products inventory, which lists consumer products that purport to use nanomaterials or nanotechnologies but are unverified. According to PEN, "The inventory now includes 1,317 products, and has grown 521% since March 2006. Europe has overtaken East Asia as the second greatest contributor of manufacturer-identified nanotechnology-enabled consumer products, though the United States still contributes the most" (WWCS 2012).

A number of environmental and consumer nongovernmental organizations (NGOs) have weighed in on nanotechnology risks over the last decade, with varying degrees of accuracy and scientific foundation. NGOs have centered their concerns in two primary areas: (1) ethical, legal, and social issues (such as public policy, equitable development, intellectual property rights, international trade, molecular manufacturing, nonproliferation, public engagement, and sustainable development), and (2) environment, health, and safety (EHS) issues (including health risk assessment, safety within the R&D community, regulations, and corporate standards).

A coalition of advocacy organizations filed the first lawsuit over nanotechnology risks against FDA for failure to regulate nanomaterials in products it oversees, led by the International Center for Technology Assessment (ICTA) on behalf of fellow plaintiffs Friends of the Earth, Food and Water Watch, the Center for Environmental Health, the ETC Group, and the Institute for Agricultural and Trade Policy (ICTA 2011). The 2011 lawsuit follows on a 2006 petition to FDA, which is responsible for regulating medical products, food products, and cosmetics, regarding their failure to inform the public about the presence of nanomaterials in products, conduct risk assessments, and ensure their safety. While food and medical applications require premarket reviews and assessments, consumer products, such as sunscreens and cosmetics, do not require premarket approvals.

Interestingly, a bill introduced to the 112th U.S. Congress, H.R. 2359, Safe Cosmetics Act of 2011, would require FDA to ensure the safety of all cosmetics and require registration and labeling, including labeling of nanomaterials in products, among other changes that would be more in line with the EC cosmetics regulation (GPO 2011). Also of interest is that only one newspaper covered the story, while word spread via the many blogs and industry trade websites (Shaw 2011).

Many of the organizations working on nanotechnology and risk issues in 2006 and 2007—when the first edition of this book was written—are no longer providing leadership or have shifted focus to other issues. There are now many more organizations and blog writers focused on nanotechnology and risk than before, and the situation generally remains one of an emerging field, rather than a mature one. If nanotechnology growth was at a pubescent stage in the mid-2000s, my characterization now would be adolescence, not even near adulthood.

There has been significant growth, and stagnation, amid continued uncertainty about the health and environmental risks of nanomaterials, nanotechnologies, and nano-enabled products, and whether we can, as a society, harness the benefits of these materials without creating unnecessary risks. Government, with fewer resources, faces industry, with limited appetite for risk, and the public and its advocates with increased demands for safety and transparency. We've made progress over the last decade on understanding and managing risks of nanomaterials and nanotechnologies, but to safely navigate the nano-era into the twenty-first century, we must increase the proactive efforts to create safer materials, and protect public health and the environment in the process.

10.5 Summary and Conclusions

This book presents a multidisciplinary evaluation of environmental and health aspects of nanotechnology. The rapid developments in this arena mean that the information herein represents a snapshot in time. The state-of-the-science regarding nanotechnology risks is a moving target. As with any emerging issue, the regulatory landscape, the organizations involved, and current thinking inevitably will change, perhaps outdating some information presented here. Nevertheless, the adaptive approaches proposed here promise continued learning and development from past and current experiences.

The complexity of our technological world, and the rapid pace of technological evolution, demands that we pay attention and participate in efforts to evaluate and manage the risks that affect us. As new technologies develop, a crucial task is to establish processes for continued surveillance to identify and address potential risks. Only through proactive efforts to understand the health and environmental impacts can we expect to responsibly manage the potential risks from nanotechnology.

References

ANF. 2011. Asia Nano Forum newsletter. No. 12. January. http://nano-globe.biz/Newsletter/ANF/ANFNewsletterIssue12.pdf

Canadian Council of Academies. 2009. Nanotechnology: Small is Different. A Science Perspective on the Regulatory Challenges of the Nanoscale. http://www.scienceadvice.ca/uploads/eng/assessments%20and%20publications%20and%20news%20releases/nano/(2008_07_10)_report_on_nanotechnology.pdf

CEN. 2011. Technical Committee 352 Work Programme. http://www.cen.eu/cen/Sectors/TechnicalCommitteesWorkshops/CENTechnicalCommittees/Pages/default.aspx?param=508478&title=CEN/TC 352

CLF Ventures. 2012. www.clfventures.org

EC. 2011a. *Official Journal of the European Union* L 342, Vol. 54, 22 December. http://eur-lex.europa.eu/JOHtml.do?uri=OJ:L:2011:342:SOM:EN:HTML

———. 2011b. Directive 2011/65/EU of the European Parliament and of the Council of 8 June 2011 on the restriction of the use of certain hazardous substances in electrical and electronic equipment. *Official Journal of the European Union* L 174 (01/07/2011): 88–110. http://eur-lex.europa.eu/LexUriServ/LexUriServ.do?uri=OJ:L:2011:174:0088:0110:EN:PDF

———. 2011d. European Commission recommendation on the definition of nanomaterial. http://osha.europa.eu/en/news/eu-european-commission-recommendation-on-the-definition-of-nanomaterial

EFSA. 2011. Guidance on the risk assessment of the application of nanoscience and nanotechnologies in the food and feed chain. *EFSA Journal* 9(5): 2140.

EPA. 2009. TSCA inventory status of nanoscale substances: General approach. http://www.epa.gov/oppt/nano/nmsp-inventorypaper.pdf

———. 2011a. Significant new use rules: Multi-walled carbon nanotubes. Docket ID EPA-HQ-OPPT-2009-0686; FRL-8865-4. Federal Register 76(88): 26186. http://www.gpo.gov/fdsys/pkg/FR-2011-05-06/pdf/2011-11127.pdf

———. 2011b. TSCA Section 5 Premanufacturing Notice Submission, Attachment 4. EPA Docket ID EPA-HQ-OPPT-2009-0686-0002. http://www.regulations.gov/#!documentDetail;D=EPA-HQ-OPPT-2009-0686-0002

———. 2011c. Decision document: Conditional registration of HeiQ AGS-20 as a materials preservative in textiles. EPA-HQ-OPP-2009-1012-0064. http://www.regulations.gov/#!documentDetail;D=EPA-HQ-OPP-2009-1012-0064

———. 2011d. Office of Inspector General. EPA needs to manage nanomaterial risks more effectively. 12-P-0162. http://www.epa.gov/oig/reports/2012/20121229-12-P-0162.pdf

FDA. 2011. Draft guidance for industry: Considering whether an FDA-regulated product involves the application of nanotechnology: Availability. Notice by the Food and Drug Administration on Federal Register 06/14/2011. http://www.federalregister.gov/articles/2011/06/14/2011-14643/draft-guidance-for-industry-considering-whether-an-fda-regulated-product-involves-the-application-of-nanotechnology

Foresight. 2012. Foresight Nanotechnology Institute. http://Foresight.org

FSANZ. 2011. Nanotechnology and food. Food Standards Australia New Zealand. http://www.foodstandards.gov.au/consumerinformation/nanotechnologyandfoo4542.cfm

GPO. 2011. 112th Congress, 1st Session, H.R. 2359. http://www.gpo.gov/fdsys/pkg/BILLS-112hr2359ih/pdf/BILLS-112hr2359ih.pdf

Health Canada. 2011. Policy statement on Health Canada's working definition for nanomaterial. http://www.hc-sc.gc.ca/sr-sr/pubs/nano/pol-eng.php

ICTA. 2011. Consumer safety groups file first lawsuit on risks of nanotechnology. International Center for Technology Assessment. http://www.icta.org/files/2011/12/Nano-PR.pdf

ISO. 2012. ISO/TR 13014:2012. Nanotechnologies: Guidance on physico-chemical characterization of engineered nanoscale materials for toxicologic assessment. http://www.iso.org/iso/iso_catalogue/catalogue_tc/catalogue_detail.htm?csnumber=52334

Lovy, H. 2005. When NanoPants attack. *Wired* 10 June. http://www.wired.com/medtech/health/news/2005/06/67626?currentPage=all

METI. 2011. Establishing the "Committee on Safety Management for Nanomaterials." http://www.meti.go.jp/english/press/2011/1130_02.html

Nakanishi, J. 2011. Risk Assessment of Manufactured Nanomaterials Final Report, "Approaches" – Overview of Approaches and Results Executive Summaries of – Carbon Nanotubes (CNT) – Fullerene (C60) – and –Titanium Dioxide . New Energy and Industrial Technology Development Organization of Japan.

Nanorelease. 2012. NanoRelease consumer products. http://www.ilsi.org/ResearchFoundation/Pages/NanoRelease1.aspx

NICNAS. 2011. Guidance on new chemical requirements for notification of industrial nanomaterials. Sydney, Australia: Australian Government, National Industrial Chemicals Notification and Assessment Scheme. http://www.nicnas.gov.au/Current_Issues/Nanotechnology/Guidance%20on%20New%20Chemical%20Requirements%20for%20Notification%20of%20Industrial%20Nanomaterials.pdf

NNI 2011. National Nanotechnology Initiative Environmental Health and Safety Research Strategy. http://nano.gov/sites/default/files/pub_resource/nni_2011_ehs_research_strategy.pdf

Nowack, B., H. F. Krug, and M. Height. 2011. 120 Years of Nanosilver History: Implications for Policy Makers. *Environ. Sci. Technol.* 45: 1177–1183.

NRDC. 2010. Comments from the Natural Resources Defense Council on the proposed conditional registration of a pesticide product HeiQ AGS-20, containing nanosilver. EPA-HQ-OPP-2009-1012. Natural Resources Defense Council. http://switchboard.nrdc.org/blogs/mwu/NRDC%20nanosilver%20CR%20Docket%20ID%20EPA-HQ-OPP-2009-1012.pdf

———. 2012. NRDC Switchboard Blog posting by Jennifer Sass. January 26. NRDC files lawsuit blocking untested nanosilver pesticide from clothing. http://switchboard.nrdc.org/cgi-bin/mt/mt-search.cgi?tag=nanosilver&limit=20

OECD. 2007. Environment directorate joint meeting of the Chemicals Committee and the Working Party on chemicals, pesticides and biotechnology. Current developments/activities on the safety of manufactured nanomaterials. Berlin, Germany. 25–27 April. ENV/JM/MONO(2007)16. http://www.oecd.org/officialdocuments/displaydocumentpdf/?cote=env/jm/mono%282007%2916&doclanguage=en

———. 2010a. Preliminary guidance notes on sample preparation and dosimetry for the testing of manufactured nanomaterials. ENV/JM/MONO(2010)25. http://www.oecd.org/officialdocuments/displaydocumentpdf/?cote=env/jm/mono%282010%2925&doclanguage=en

———. 2010b. OECD Environment, Health and Safety Publications Series on the Safety of Manufactured Nanomaterials, No. 27. List of manufactured nanomaterials and list of endpoints for phase one of the sponsorship programme for the testing of manufactured nanomaterials: Revision. ENV/JM/MONO(2010)46. http://www.oecd.org/officialdocuments/displaydocumentpdf/?cote=env/jm/mono%282010%2946&doclanguage=en

———. 2011. Current developments/activities on the safety of manufactured nanomaterials. Tour de Table at the 8th Meeting of the Working Party on Manufactured Nanomaterials. Paris, France. 16–18 March. ENV/JM/MONO(2011)12. http://www.oecd.org/officialdocuments/displaydocumentpdf/?cote=env/jm/mono%282011%2912&doclanguage=en

OMB. 2011. Policy principles for the U.S. decision-making concerning regulation and oversight of applications of nanotechnology and nanomaterials. http://www.whitehouse.gov/sites/default/files/omb/inforeg/for-agencies/nanotechnology-regulation-and-oversight-principles.pdf

Shatkin, J. A., and W. North. 2010. Perspectives on risks of nanomaterials and nanotechnologies: Advancing the science. *Risk Analysis* 30(11): 1627–33.

Shaw, G. 2011. FDA sued on nano. *New Haven Independent*, December 21. http://www.newhavenindependent.org/index.php/archives/entry/advocacy_groups_sue_to_force_fd/

SOT. 2007. Communiqué on-line. Society of Toxicology. http://www.toxicology.org/AI/PUB/si07/si07_nano.asp

SRA. 2012. Society for Risk Analysis Risk Newsletter. http://www.sra.org/newsletter/RISK.V32.N1.pdf

Stone, V., H. Johnston, and R. P. F. Schins. 2009. Review article: Development of in vitro systems for nanotoxicology: Methodological considerations. *Crit. Rev. Toxicol.* 39(7): 613–26.

WWCS. 2012. Project on emerging nanotechnologies. http://www.nanotechproject.org/

Further Reading

ABA. 2007. Section nanotechnology project. Section of Environment, Energy, and Resources. http://www.americanbar.org/groups/environment_energy_resources.html

Acción Ecológica, et al. 2007. Principles for the oversight of nanotechnologies and nanomaterials. http://www.foe.org/pdf/Nanotech_Principles.pdf

Air Products, Altair Nanotechnologies, American Chemistry Council, Arkema, BASF, Bayer, Degussa, et al. 2007. Letter to the Senate and House Appropriations Committee. http://www.environmentaldefense.org/documents/6015_Approps_2007NASLetter.pdf

Ata, M. 2007. Nanotechnology: Toward innovation and a society of sustainable development. The challenge for a new methodology of technology development. *AIST Today* 2007 (23). http://www.aist.go.jp/aist_e/aist_today/2007_23/nanotec/nanotec_09.html

CEN. 2007. Building up nanotech research. *Chem. Eng. News* 85(15): 15–21.

Chinese Academy of Sciences. 2007. Studies kick off on bio-safety of artificial nano-materials. http://english.cas.ac.cn/eng2003/news/detailnewsb.asp?infono=26361

Cordis. 2007. Seventh Research Framework Programme (FP7). http://cordis.europa.eu/fp7/home_en.html

CTBA. 2007a. Mission and purpose statements. Converging Technologies Bar Association. http://www.convergingtechnologies.org/aboutctba.asp

———. 2007b. Welcome message. Converging Technologies Bar Association. http://www.convergingtechnologies.org/default2.asp

Davies, J. C. 2006. Managing the effects of nanotechnology. Project on Emerging Nanotechnologies. Washington, DC. http://www.nanotechproject.org/publications/archive/managing_effects_nanotechnology/

———. 2007. EPA and nanotechnology: Oversight for the 21st century. Project on Emerging Nanotechnologies. Washington, DC. http://www.nanotechproject.org/publications/archive/epa_nanotechnology_oversight_for_21st/

Environment Canada. 2007. New Substances Program Advisory Note 2007-06. Requirements for nanomaterials under the New Substances Notification Regulations (Chemicals and Polymers)m http://www.ec.gc.ca/subsnouvelles-newsubs/default.asp?lang=En&n=3C32F773-1

EC. 2011c. Regulation (EU) No. 1169/2011 of the European Parliament and of the Council of 25 October 2011. *Official Journal of the European Union* L 304 (22/11/2011): 18–63. http://eur-lex.europa.eu/LexUriServ/LexUriServ.do?uri=OJ:L:2011:304:0018:0063:EN:PDF

ICON. 2007. International Council on Nanotechnology. http://icon.rice.edu/

Industry Australia. 2007. National nanotechnology strategy. http://www.innovation.gov.au/INDUSTRY/NANOTECHNOLOGY/NATIONALENABLINGTECHNOLOGIESSTRATEGY/Pages/NationalNanotechnologyStrategy.aspx

IRGC. 2006. About IRGC. http://www.irgc.org/spip/spip.php?page=irgc&id_rubrique=3

Maynard A. D., R. J. Aitken, T. Butz, V. Colvin, K. Donaldson, G. Oberdörster, M. A. Philbert, et al. 2006. Safe handling of nanotechnology. *Nature* 444:267–69.

Nanoreg News. 2007. Nanoreg report. http://www.nanoregnews.com/

NEHI. 2007. Prioritization of environmental, health, and safety research needs for engineered nanoscale materials: An interim document for public comment. National Nanotechnology Coordination Office. http://www.nano.gov/sites/default/files/pub_resource/nni_ehs_research_needs.pdf?q=NNI_EHS_research_needs.pdf&q=Prioritization_EHS_Research_Needs_Engineered_Nanoscale_Materials.pdf

NNI. 2007. Public shares views on EHS research needs for engineered nanoscale materials. http://nano.gov/node/647

NRC. 2006. *A Matter of Size: Triennial Review of the National Nanotechnology Initiative.* Washington, DC: National Academy Press.

Index